The Uninformed States of America

How Scientific Illiteracy Undermines Our Progress

A Short Introduction by K.R. Stonebridge

The Uninformed States of America:
How Scientific Illiteracy Undermines Our Progress
  ISBN: 9798328049825

  Copyright © 2024 Brevity Books

All rights reserved. No part of this publication may be reproduced, stored on a retrieval system, or transmitted, in any form or by any means, electronic, mechanical, photocopying, microfilming, recording or otherwise, without written permission from the publisher.

Printed in the United States of America

Contents

Introduction

The Uninformed States of America: How Scientific Illiteracy Undermines Our Progress..........9

Thesis Statement: Scientific Illiteracy Poses a Significant Threat to America's Progress and Well-being....11

Part I: The State of Scientific Illiteracy in America

Chapter 1: The Extent of the Problem

Statistics on Scientific Knowledge Among Americans........15

Comparison with Other Developed Nations..........18

Historical Trends and Factors Contributing to the Current Situation..........21

Chapter 2: The Educational System's Role

Inadequacies in K-12 Science Education..........25

Challenges Faced by Educators and Students..........28

The Impact of Politicization and Controversies on Science Education..........31

Chapter 3:
Media, Misinformation, and Pseudoscience

The Role of Media in Shaping Public Understanding of Science 35

The Spread of Misinformation and Pseudoscience 38

The Influence of Social Media and the Internet 42

Part II:
The Consequences of Scientific Illiteracy

Chapter 4:
Public Health and Personal Well-being

Misconceptions about Diseases, Vaccines, and Treatments 47

The Impact on Personal Health Decisions and Public Health Policies 51

Case Studies: The Anti-Vaccination Movement and the COVID-19 Response 55

Chapter 5:
Environmental Issues and Climate Change

Misunderstanding of Environmental Science and Climate Change 59

The Impact on Individual Actions and Support for Environmental Policies 63

Consequences for the Planet and Future Generations 66

Chapter 6:
Technological Advancement and Economic Competitiveness

The Importance of Science and Technology for Innovation and Economic Growth 71

The Impact of Scientific Illiteracy on America's Workforce and Competitiveness 75

Case Studies: Declining Interest in STEM Fields and Brain Drain 79

Chapter 7:
Political Decision-Making and Democracy

The Role of Science in Informing Policy Decisions 84

The Impact of Scientific Illiteracy on Voter Choices and Political Discourse 88

Consequences for Evidence-Based Policymaking and Democracy 91

Part III:
Addressing the Problem

Chapter 8:
Improving Science Education

Strategies for Enhancing K-12 Science Curricula and Teaching Methods 97

The Role of Universities and Informal Science Education ... 101

Successful Programs and Initiatives 105

Chapter 9:
Promoting Science Communication and Public Engagement

The Importance of Effective Science Communication 110
Strategies for Scientists, Journalists, and Institutions 114
Successful Examples of Public Engagement with Science 118

Chapter 10:
Building a Scientifically Literate Society

The Benefits of a Scientifically Literate Population 123
The Role of Individuals, Communities, and Policymakers 127
A Vision for the Future and a Call to Action 130

Conclusion

Recap of the Book's Main Points 135
The Urgency of Addressing Scientific Illiteracy in America 138
The Potential for Positive Change and Progress 141

Appendices

Resources for Further Reading and Engagement 144
Glossary of Key Scientific Terms and Concepts 148

Introduction

The Uninformed States of America: How Scientific Illiteracy Undermines Our Progress

In an age where science and technology play an increasingly crucial role in our daily lives, the United States finds itself grappling with a troubling phenomenon: widespread scientific illiteracy. This lack of understanding and appreciation for scientific principles and methods has far-reaching consequences, impacting not only individuals but also society as a whole. Scientific illiteracy can be defined as the inability to comprehend and apply basic scientific concepts, coupled with a lack of awareness about the scientific process and its significance in decision-making [1].

The prevalence of scientific illiteracy in the United States is alarming. A 2018 survey conducted by the Pew Research Center revealed that only 52% of American adults could correctly answer a series of basic science questions [2]. This means that nearly half of the adult population struggles to grasp fundamental scientific ideas. Moreover, the survey found that scientific knowledge varies significantly based on factors such as education level, age, and political affiliation, highlighting the complex nature of the issue.

The consequences of scientific illiteracy are manifold. On an individual level, it can lead to poor decision-making regarding personal health, nutrition, and environmental choices. For example, a lack of understanding about the importance of vaccinations has contributed to the resurgence of preventable diseases in some communities [3]. Furthermore, scientific illiteracy can make individuals more susceptible to misinformation and pseudoscience, which can have detrimental effects on their well-being and that of those around them.

At a societal level, scientific illiteracy can hinder progress and innovation. In an increasingly competitive global economy, a scientifically literate workforce is essential for driving technological advancements and maintaining economic competitiveness [4]. When a significant portion of the population lacks the necessary scientific skills and knowledge, it becomes challenging to foster a culture of innovation and adapt to the rapidly changing demands of the 21st century.

Moreover, scientific illiteracy can undermine the democratic process by impairing the public's ability to make informed decisions on science-related policy issues. From climate change to biotechnology, many of the most pressing challenges facing society today require a solid understanding of scientific principles [5]. When voters and policymakers lack this understanding, it becomes difficult to engage in meaningful discourse and make evidence-based decisions that prioritize the long-term well-being of the nation.

The factors contributing to scientific illiteracy in the United States are complex and multifaceted. One significant factor is the inadequacy of science education in many K-12 schools. A report by the National Academy of Sciences found that science education in the United States often lacks depth, coherence, and rigor, leaving students ill-prepared to apply scientific thinking in their lives [6]. Additionally, the politicization of science and the spread of misinformation through various media channels have further contributed to the erosion of scientific understanding among the public.

Addressing the problem of scientific illiteracy requires a concerted effort from educators, policymakers, scientists, and the media. Improving science education, promoting effective science communication, and fostering a culture that values scientific thinking are all essential steps in building a more scientifically literate society. By doing so, we can ensure that the United States is well-positioned to tackle the challenges of the future and maintain its role as a global leader in science and innovation.

References

1. National Academies of Sciences, Engineering, and Medicine. (2016). *Science literacy: Concepts, contexts, and consequences*. National Academies Press.
2. Pew Research Center. (2018). *What Americans know about science*.
3. Omer, S. B., Salmon, D. A., Orenstein, W. A., deHart, M. P., & Halsey, N. (2009). Vaccine refusal, mandatory immunization, and the risks of vaccine-preventable diseases. *New England Journal of Medicine, 360*(19), 1981-1988.
4. National Science Board. (2018). *Science and engineering indicators 2018*. National Science Foundation.
5. Miller, J. D. (2004). Public understanding of, and attitudes toward, scientific research: What we know and what we need to know. *Public Understanding of Science, 13*(3), 273-294.
6. National Research Council. (2007). *Taking science to school: Learning and teaching science in grades K-8*. National Academies Press.

Thesis Statement: Scientific Illiteracy Poses a Significant Threat to America's Progress and Well-being

The United States has long prided itself on being a leader in scientific and technological advancement, with innovations that have transformed the world and improved the lives of countless individuals. However, the pervasive lack of scientific understanding among the American public threatens to undermine the very foundation upon which this progress has been built. Scientific illiteracy, characterized by a limited grasp of basic scientific concepts and a lack of appreciation for the scientific process, has far-reaching consequences that extend beyond the realm of academia and into the very fabric of our society [1].

The impact of scientific illiteracy on America's progress and well-being is multifaceted and profound. In an era where science and technology are increasingly intertwined with every aspect of our lives, from healthcare and education to public policy and economic development, a scientifically illiterate population is ill-equipped to make informed decisions and adapt to the challenges of the 21st century [2].

One of the most pressing concerns related to scientific illiteracy is its impact on public health. When individuals lack a basic understanding of scientific principles, they are more likely to fall prey to misinformation and pseudoscience, which can lead to dangerous health choices [3]. The anti-vaccination movement, for example, has gained traction in recent years, fueled in part by a lack of understanding about the safety and effectiveness of vaccines. This has led to outbreaks of preventable diseases, such as measles, in communities with low vaccination rates [4]. Moreover, scientific illiteracy can hinder the public's ability to comprehend and follow evidence-based health guidelines, as evidenced by the challenges faced in combating the COVID-19 pandemic [5].

Scientific illiteracy also poses a threat to America's economic competitiveness and technological advancement. In an increasingly globalized and knowledge-based economy, a scientifically literate workforce is essential for driving innovation and maintaining a competitive edge [6]. When a significant portion of the population lacks the necessary scientific skills and knowledge, it becomes difficult to cultivate a robust talent pipeline in science, technology, engineering, and mathematics (STEM) fields. This, in turn, can lead to a shortage of qualified professionals in key industries, hindering economic growth and limiting America's ability to tackle complex challenges such as climate change, energy security, and public health crises [7].

Furthermore, scientific illiteracy can undermine the democratic process by impairing the public's ability to engage in informed decision-making on science-related policy issues. Many of the most pressing challenges facing society today, from climate change to genetic engineering, require a nuanced understanding of scientific principles [8]. When voters and policymakers lack this understanding, it becomes challenging to have productive public discourse and make evidence-based decisions that prioritize the long-term well-being of the nation. This can lead to the implementation of policies that are not grounded in scientific evidence, potentially causing harm to individuals, communities, and the environment [9].

Addressing the problem of scientific illiteracy requires a multifaceted approach that involves educators, policymakers, scientists, and the media. Improving science education in K-12 schools is a critical step in fostering a scientifically literate population. This involves not only teaching scientific concepts and skills but also cultivating an appreciation for the scientific process and its role in evidence-based decision-making [10]. Additionally, promoting effective science communication and public engagement is essential for bridging the gap between the scientific community and the general public. By making scientific knowledge more accessible and relatable, we can help individuals develop a greater understanding and appreciation for science and its importance in their lives [11].

In conclusion, scientific illiteracy poses a significant threat to America's progress and well-being. It undermines public health, hinders economic competitiveness, and impairs the democratic process. Addressing this issue requires a sustained and collaborative effort from all sectors of society. By prioritizing science education, promoting effective science communication, and fostering a culture that values scientific thinking, we can build a more scientifically literate society that is better equipped to tackle the challenges of the 21st century and ensure a brighter future for all.

References

1. National Academies of Sciences, Engineering, and Medicine. (2016). *Science literacy: Concepts, contexts, and consequences*. National Academies Press.
2. Miller, J. D. (2004). Public understanding of, and attitudes toward, scientific research: What we know and what we need to know. *Public Understanding of Science, 13*(3), 273-294.
3. Scheufele, D. A. (2013). Communicating science in social settings. *Proceedings of the National Academy of Sciences, 110*(Supplement 3), 14040-14047.
4. Omer, S. B., Salmon, D. A., Orenstein, W. A., deHart, M. P., & Halsey, N. (2009). Vaccine refusal, mandatory immunization, and the risks of vaccine-preventable diseases. *New England Journal of Medicine, 360*(19), 1981-1988.
5. Paakkari, L., & Okan, O. (2020). COVID-19: Health literacy is an underestimated problem. *The Lancet Public Health, 5*(5), e249-e250.
6. National Science Board. (2018). *Science and engineering indicators 2018*. National Science Foundation.
7. Holdren, J. P., Marrett, C., & Suresh, S. (2013). Federal Science, Technology, Engineering, and Mathematics (STEM) education 5-year strategic plan. *National Science and Technology Council: Committee on STEM Education*.
8. Dietz, T. (2013). Bringing values and deliberation to science communication. *Proceedings of the National Academy of Sciences, 110*(Supplement 3), 14081-14087.
9. Druckman, J. N. (2015). Communicating policy-relevant science. *PS: Political Science & Politics, 48*(S1), 58-69.
10. National Research Council. (2012). *A framework for K-12 science education: Practices, crosscutting concepts, and core ideas*. National Academies Press.
11. Fischhoff, B. (2013). The sciences of science communication. *Proceedings of the National Academy of Sciences, 110*(Supplement 3), 14033-14039.

Part I:
The State of Scientific Illiteracy in America

Chapter 1: The Extent of the Problem

Statistics on Scientific Knowledge Among Americans

The state of scientific literacy in the United States has been a topic of concern for educators, policymakers, and researchers alike. To gauge the extent of the problem, it is essential to examine the available data on Americans' understanding of basic scientific concepts and their ability to apply this knowledge in real-world situations.

One of the most comprehensive assessments of scientific knowledge among Americans is the National Science Foundation's (NSF) Science and Engineering Indicators report, which is published every two years [1]. The report includes data from the National Survey of Adult Literacy, which measures the scientific literacy of U.S. adults across three dimensions: factual knowledge, understanding of scientific processes, and the ability to apply scientific information in everyday life [2].

According to the 2018 Science and Engineering Indicators report, only 22% of American adults were considered scientifically literate, meaning they could correctly answer a series of questions designed to assess their understanding of basic scientific concepts and processes [1]. This figure has remained relatively stable since the early 2000s, indicating a persistent lack of scientific understanding among a significant portion of the population.

The report also found that scientific literacy varies significantly based on factors such as education level, age, and socioeconomic status. For example, adults with a bachelor's degree or higher were more than twice as likely to be scientifically literate compared to those with only a high school diploma (44% vs. 17%) [1]. Similarly, younger adults (aged 18-34) were more likely to be scientifically literate than older adults (aged 65 and above) (27% vs. 16%) [1].

Another important source of data on scientific knowledge among Americans is the Pew Research Center's Science Knowledge Quiz, which has been administered to representative samples of U.S. adults on multiple occasions since 2005 [3]. The quiz consists of a series of multiple-choice questions covering a range of scientific topics, including basic physics, genetics, and earth sciences.

In the most recent iteration of the quiz, conducted in 2019, the average score was 6.7 out of 11 questions answered correctly [3]. While this represents a slight improvement from previous years, it still indicates a significant gap in scientific understanding among the general public. Notably, the quiz revealed that many Americans struggle with concepts related to scientific inquiry and the nature of scientific evidence. For example, only 52% of respondents correctly identified that scientific hypotheses must be testable and falsifiable [3].

The Pew Research Center's quiz also highlighted the role of education in shaping scientific knowledge. Adults with a postgraduate degree scored significantly higher on the quiz compared to those with only a high school diploma or less (8.8 vs. 5.3 correct answers, on average) [3]. This finding underscores the importance of access to quality science education in fostering a scientifically literate population.

In addition to these national surveys, numerous studies have investigated scientific knowledge and understanding in specific domains, such as health and environmental science. For example, a 2017 study published in the Journal of Health Communication found that only 12% of American adults had a proficient level of health literacy, meaning they could effectively obtain, process, and understand basic health information and services needed to make appropriate health decisions [4]. This lack of health literacy can have serious consequences, as it can lead to poor health outcomes, increased healthcare costs, and a reduced ability to engage in preventive health behaviors [5].

Similarly, research has shown that many Americans struggle to understand key concepts related to climate change and other environmental issues. A 2019 survey conducted by the Yale Program on Climate Change Communication found that while a majority of Americans believe climate change is happening (66%), only 54% believe it is caused primarily by human activities [6]. Moreover, the survey revealed significant gaps in understanding of the scientific consensus on climate change, with only 19% of respondents correctly identifying that 91-100% of climate scientists agree that human-caused global warming is occurring [6].

These statistics paint a troubling picture of the state of scientific knowledge among Americans. While there have been some modest improvements in recent years, the overall level of scientific literacy remains low, with significant disparities based on factors such as education, age, and socioeconomic status. This lack of scientific understanding has far-reaching consequences, impacting individuals' ability to make informed decisions about their health, the environment, and other important issues.

Addressing this problem will require a concerted effort to improve science education at all levels, from K-12 schools to adult learning programs. It will also require effective communication strategies to help bridge the gap between the scientific community and the general public, making scientific knowledge more accessible and relatable to everyday life. By prioritizing these efforts, we can work towards building a more scientifically literate society that is better equipped to tackle the complex challenges of the 21st century.

References

1. National Science Board. (2018). *Science and Engineering Indicators 2018*. National Science Foundation.
2. Kutner, M., Greenberg, E., Jin, Y., & Paulsen, C. (2006). *The Health Literacy of America's Adults: Results from the 2003 National Assessment of Adult Literacy*. National Center for Education Statistics.
3. Pew Research Center. (2019). *What Americans Know About Science*.
4. Kutner, M., Greenberg, E., Jin, Y., Paulsen, C., & White, S. (2006). *The Health Literacy of America's Adults: Results from the 2003 National Assessment of Adult Literacy*. National Center for Education Statistics.

5. Berkman, N. D., Sheridan, S. L., Donahue, K. E., Halpern, D. J., & Crotty, K. (2011). Low health literacy and health outcomes: An updated systematic review. *Annals of Internal Medicine, 155*(2), 97-107.
6. Leiserowitz, A., Maibach, E., Rosenthal, S., Kotcher, J., Bergquist, P., Ballew, M., Goldberg, M., & Gustafson, A. (2019). *Climate Change in the American Mind: November 2019*. Yale Program on Climate Change Communication.

Comparison with Other Developed Nations

To fully understand the extent of scientific illiteracy in the United States, it is essential to examine how the nation fares in comparison to other developed countries. International assessments and surveys provide valuable insights into the relative strengths and weaknesses of American scientific knowledge and underscore the need for targeted efforts to improve scientific literacy.

One of the most comprehensive international assessments of scientific literacy is the Programme for International Student Assessment (PISA), conducted by the Organisation for Economic Co-operation and Development (OECD) [1]. PISA measures the performance of 15-year-old students in reading, mathematics, and science across more than 70 countries and economies. The assessment focuses on students' ability to apply their knowledge and skills to real-world situations, providing a valuable indicator of their preparedness for future challenges.

In the most recent PISA assessment, conducted in 2018, the United States ranked 18th out of 79 participating countries and economies in science performance, with an average score of 502 points [1]. While this score is slightly above the OECD average of 489 points, it falls well behind top-performing countries such as China (590 points), Singapore (551 points), and Estonia (530 points) [1]. Moreover, the U.S. score has remained relatively stagnant since the first PISA assessment in 2000, indicating a lack of significant progress in improving scientific literacy among American students.

The PISA results also reveal significant disparities in scientific performance within the United States, with students from disadvantaged socioeconomic backgrounds scoring significantly lower than their more advantaged peers [1]. This achievement gap is more pronounced in the United States compared to many other developed nations, suggesting that issues of equity and access to quality science education are particularly pressing in the American context.

Another important international comparison of scientific literacy comes from the Trends in International Mathematics and Science Study (TIMSS), conducted by the International Association for the Evaluation of Educational Achievement (IEA) [2]. TIMSS assesses the mathematics and science knowledge and skills of fourth and eighth-grade students across more than 60 countries and benchmarking entities.

In the 2019 TIMSS assessment, U.S. eighth-grade students ranked 11th out of 39 participating countries in science achievement, with an average score of 522 points [2]. While this represents a significant improvement from the U.S. ranking of 23rd in the 2015 TIMSS assessment, it still places the nation behind several other developed countries, such as Singapore (608 points), Japan (570 points), and the Russian Federation (543 points) [2].

The TIMSS results also highlight the importance of early science education in fostering scientific literacy. In the fourth-grade assessment, U.S. students ranked 10th out of 58 participating countries, with an average score of 539 points [2]. However, this relatively strong performance in the early grades does not necessarily translate into sustained scientific literacy in later years, as evidenced by the comparatively weaker performance of U.S. students in the eighth-grade assessment and in the PISA assessment of 15-year-olds.

Beyond student assessments, international surveys of adult scientific literacy also provide valuable insights into how the United States compares to other developed nations. The Programme for the International Assessment of Adult Competencies (PIAAC), conducted by the OECD, assesses the skills and competencies of adults aged 16-65 across more than 40 countries [3]. While PIAAC does not specifically test scientific literacy, it does measure key skills that are essential for understanding and applying scientific information, such as literacy, numeracy, and problem-solving in technology-rich environments.

In the most recent PIAAC assessment, conducted in 2017, the United States ranked 19th out of 39 participating countries in literacy, with an average score of 271 points (compared to an OECD average of 268 points) [3]. In numeracy, the United States ranked 24th, with an average score of 255 points (compared to an OECD average of 263 points) [3]. These results suggest that American adults, on average, have relatively weak skills in key areas that are essential for scientific literacy, which may limit their ability to effectively engage with scientific information and make informed decisions.

The international comparisons presented here paint a sobering picture of the state of scientific literacy in the United States. While the nation performs slightly above average in some assessments, it consistently lags behind top-performing countries, particularly in Asia and Europe. Moreover, the significant disparities in scientific performance within the United States, particularly along socioeconomic lines, underscore the need for targeted efforts to improve equity and access to quality science education.

Addressing these challenges will require a multifaceted approach that involves strengthening science education at all levels, from early childhood through adulthood. This may include investing in teacher training and professional development, updating curricula to emphasize real-world applications of scientific knowledge, and providing resources and support to schools and communities that serve disadvantaged populations. Additionally, efforts to promote public engagement with science and to improve science communication and outreach can help foster a culture of scientific literacy and critical thinking among American adults.

By learning from the successes of other developed nations and prioritizing efforts to improve scientific literacy, the United States can work towards building a more scientifically engaged and informed society. This, in turn, can help the nation better address the complex challenges of the 21st century, from public health crises to environmental sustainability, and maintain its position as a global leader in science and innovation.

References
1. OECD. (2019). *PISA 2018 Results (Volume I): What Students Know and Can Do*. OECD Publishing.
2. Mullis, I. V. S., Martin, M. O., Foy, P., Kelly, D. L., & Fishbein, B. (2020). *TIMSS 2019 International Results in Mathematics and Science*. Boston College, TIMSS & PIRLS International Study Center.
3. OECD. (2019). *Skills Matter: Additional Results from the Survey of Adult Skills*. OECD Skills Studies. OECD Publishing.

Historical Trends and Factors Contributing to the Current Situation

The current state of scientific illiteracy in the United States is not a sudden phenomenon but rather the result of a complex interplay of historical trends and societal factors. To fully understand the roots of this problem, it is necessary to examine the evolution of science education, public attitudes towards science, and the broader social and political context in which scientific knowledge is produced and disseminated.

One important historical trend to consider is the changing nature of science education in the United States. In the early 20th century, science education was primarily focused on preparing students for specific vocations, such as agriculture or engineering [1]. It was not until the launch of the Soviet satellite Sputnik in 1957 that the United States began to place a greater emphasis on science education as a means of bolstering national security and technological superiority [2].

The post-Sputnik era saw a surge of federal funding for science education and research, as well as the development of new curricula and teaching methods designed to promote scientific literacy and critical thinking [2]. However, these efforts were not always sustained or evenly distributed, and many students, particularly those from disadvantaged backgrounds, continued to receive inadequate or low-quality science education.

In the latter half of the 20th century, the United States also experienced significant shifts in public attitudes towards science. The environmental movement of the 1960s and 1970s, fueled by concerns about pollution, resource depletion, and the health impacts of industrial chemicals, led to a growing public awareness of the importance of scientific research in addressing societal challenges [3]. At the same time, however, this period also saw the emergence of a more skeptical and critical view of science, as some began to question the authority of scientific experts and the motivations behind scientific research [4].

The rise of the "culture wars" in the 1980s and 1990s further polarized public attitudes towards science, particularly around issues such as evolution, climate change, and stem cell research [5]. As these debates became increasingly politicized, many Americans began to view science through a partisan lens, with some seeing it as a tool of liberal elites and others embracing it as a means of advancing social and economic progress [6].

The advent of the internet and social media in the early 21st century has added another layer of complexity to the landscape of scientific literacy. On the one hand, these technologies have made it easier than ever for individuals to access scientific information and engage with the scientific community [7]. On the other hand, they have also facilitated the rapid spread of misinformation and pseudoscience, as well as the formation of echo chambers in which individuals are exposed only to information that confirms their pre-existing beliefs [8].

The COVID-19 pandemic has brought many of these historical trends and factors into sharp relief. The public health crisis has underscored the critical importance of scientific research and evidence-based decision-making, while also exposing deep fault lines in public trust and understanding of science [9]. The politicization of issues such as mask-wearing and vaccine development, coupled with the proliferation of conspiracy theories and false information online, has further eroded scientific literacy and undermined efforts to mount an effective response to the pandemic [10].

Looking beyond the immediate crisis, it is clear that addressing scientific illiteracy in the United States will require a sustained and multifaceted effort that takes into account the complex historical and societal factors that have contributed to the current situation. This may involve reinvesting in science education at all levels, from K-12 schools to adult learning programs, and developing new curricula and teaching methods that emphasize critical thinking, data analysis, and real-world applications of scientific knowledge.

It will also require building public trust in science and scientific institutions, by promoting transparency, accountability, and ethical conduct in scientific research, and by fostering greater engagement between the scientific community and the broader public. This may involve developing new models of science communication and outreach, such as citizen science initiatives, science cafes, and online platforms for public participation in scientific research [11].

Finally, addressing scientific illiteracy will require confronting the deeper social, political, and economic factors that have contributed to the erosion of public trust and understanding of science. This may involve efforts to reduce political polarization, combat misinformation and disinformation, and address issues of inequality and social justice that have historically limited access to high-quality science education and scientific careers.

By taking a holistic and historical view of the problem of scientific illiteracy, and by developing comprehensive strategies to address its root causes, the United States can work towards building a more scientifically literate and engaged society. This, in turn, can help the nation better address the complex challenges of the 21st century, from climate change and public health to technological innovation and economic competitiveness.

References

1. Rudolph, J. L. (2002). *Scientists in the classroom: The cold war reconstruction of American science education*. Palgrave Macmillan.
2. Wissehr, C., Concannon, J., & Barrow, L. H. (2011). Looking back at the Sputnik era and its impact on science education. *School Science and Mathematics, 111*(7), 368-375.
3. Lear, L. J. (1997). *Rachel Carson: Witness for nature*. Henry Holt and Company.
4. Lewenstein, B. V. (1992). The meaning of 'public understanding of science' in the United States after World War II. *Public Understanding of Science, 1*(1), 45-68.
5. Mooney, C. (2005). *The Republican war on science*. Basic Books.
6. Gauchat, G. (2012). Politicization of science in the public sphere: A study of public trust in the United States, 1974 to 2010. *American Sociological Review, 77*(2), 167-187.
7. Brossard, D., & Scheufele, D. A. (2013). Science, new media, and the public. *Science, 339*(6115), 40-41.
8. Del Vicario, M., Bessi, A., Zollo, F., Petroni, F., Scala, A., Caldarelli, G., Stanley, H. E., & Quattrociocchi, W. (2016). The spreading of misinformation online. *Proceedings of the National Academy of Sciences, 113*(3), 554-559.
9. Kreps, S., & Kriner, D. (2020). Model uncertainty, political contestation, and public trust in science: Evidence from the COVID-19 pandemic. *Science Advances, 6*(43), eabd4563.
10. Barua, Z., Barua, S., Aktar, S., Kabir, N., & Li, M. (2020). Effects of misinformation on COVID-19 individual responses and recommendations for resilience of disastrous consequences of misinformation. *Progress in Disaster Science, 8*, 100119.
11. Bonney, R., Phillips, T. B., Ballard, H. L., & Enck, J. W. (2016). Can citizen science enhance public understanding of science? *Public Understanding of Science, 25*(1), 2-16.

Chapter 2: The Educational System's Role

Inadequacies in K-12 Science Education

The foundation for scientific literacy is laid in the formative years of an individual's education, making the quality of K-12 science education a crucial factor in determining the overall scientific literacy of a nation. Unfortunately, the United States has long struggled with inadequacies in its K-12 science education system, which have contributed significantly to the current state of scientific illiteracy among American adults.

One of the most pressing issues in K-12 science education is the lack of consistent and high-quality standards across states and school districts. While the Next Generation Science Standards (NGSS), developed in 2013, have been adopted by many states, their implementation has been uneven and often insufficient [1]. Some states have chosen to develop their own standards, leading to a patchwork of science education requirements that vary widely in terms of content, depth, and rigor [2].

This lack of consistency is further compounded by inadequate funding for science education, particularly in low-income and underserved communities. Many schools struggle to provide the necessary resources, such as well-equipped laboratories, up-to-date textbooks, and qualified science teachers, to deliver high-quality science instruction [3]. This resource gap contributes to the persistent achievement gaps in science performance between students from disadvantaged backgrounds and their more affluent peers [4].

Another significant challenge in K-12 science education is the limited time and emphasis placed on science instruction, particularly in the elementary grades. A 2018 survey by the National Survey of Science and Mathematics Education found that elementary school teachers spent an average of just 18 minutes per day on science instruction, compared to 57 minutes on reading and 54 minutes on mathematics [5]. This lack of early exposure to science can have lasting impacts on students' interest and confidence in the subject, as well as their preparedness for more advanced coursework in middle and high school [6].

The quality and effectiveness of science instruction in K-12 classrooms is also hindered by a shortage of qualified and experienced science teachers. Many science teachers, particularly at the elementary level, lack strong backgrounds in science content and pedagogy, which can limit their ability to engage students and deliver rigorous, inquiry-based instruction [7]. Additionally, high rates of teacher turnover, especially in high-needs schools, can disrupt the continuity of science instruction and make it difficult for students to build a strong foundation in the subject [8].

The inadequacies in K-12 science education are further exacerbated by the often narrow and test-driven focus of science curricula. In many schools, science instruction is heavily geared towards preparing students for standardized tests, rather than fostering deep understanding, critical thinking, and real-world application of scientific knowledge [9]. This emphasis on rote memorization and recall of facts can discourage students from pursuing further study in science and can contribute to the development of misconceptions and lack of interest in the subject [10].

Efforts to address these inadequacies in K-12 science education have often been piecemeal and insufficiently funded. While programs like the National Science Foundation's Math and Science Partnership have provided support for teacher professional development and curriculum improvement, their reach and impact have been limited by budgetary constraints and competing priorities [11]. Similarly, initiatives to promote equity and diversity in science education, such as the National Science Foundation's Broadening Participation in STEM programs, have made some progress in increasing access and representation, but much work remains to be done [12].

Improving the quality and effectiveness of K-12 science education will require a sustained and comprehensive effort that addresses the multiple factors contributing to the current inadequacies. This may involve developing and implementing consistent, high-quality science standards across all states and school districts, coupled with sufficient funding and resources to support their effective implementation. It will also require investing in the recruitment, preparation, and retention of qualified and diverse science teachers, as well as providing ongoing professional development and support to help them deliver engaging, inquiry-based instruction.

Additionally, efforts to reform K-12 science education should prioritize the development of curricula and assessments that emphasize deep understanding, critical thinking, and real-world application of scientific knowledge, rather than rote memorization and test preparation. This may involve partnerships between schools, universities, and industry to provide students with authentic scientific experiences and exposure to diverse career paths in science, technology, engineering, and mathematics (STEM) fields.

Ultimately, addressing the inadequacies in K-12 science education will require a recognition of the critical importance of scientific literacy for individual and societal well-being, as well as a commitment to providing all students, regardless of background or socioeconomic status, with access to high-quality, engaging, and relevant science instruction. By investing in the improvement of K-12 science education, the United States can lay the foundation for a more scientifically literate and engaged population, better equipped to tackle the complex challenges of the 21st century.

References

1. National Research Council. (2013). *Next Generation Science Standards: For states, by states.* National Academies Press.
2. Lerner, L. S., Goodenough, U., Lynch, J., Schwartz, M., & Schwartz, R. (2012). *The state of state science standards.* Thomas B. Fordham Institute.
3. National Science Board. (2018). *Science and engineering indicators 2018.* National Science Foundation.
4. Quinn, D. M., & Cooc, N. (2015). Science achievement gaps by gender and race/ethnicity in elementary and middle school: Trends and predictors. *Educational Researcher, 44*(6), 336-346.
5. Banilower, E. R., Smith, P. S., Malzahn, K. A., Plumley, C. L., Gordon, E. M., & Hayes, M. L. (2018). *Report of the 2018 NSSME+.* Horizon Research, Inc.
6. Morgan, P. L., Farkas, G., Hillemeier, M. M., & Maczuga, S. (2016). Science achievement gaps begin very early, persist, and are largely explained by modifiable factors. *Educational Researcher, 45*(1), 18-35.
7. Nowicki, B. L., Sullivan-Watts, B., Shim, M. K., Young, B., & Pockalny, R. (2013). Factors influencing science content accuracy in elementary inquiry science lessons. *Research in Science Education, 43*(3), 1135-1154.
8. Ingersoll, R. M., & May, H. (2012). The magnitude, destinations, and determinants of mathematics and science teacher turnover. *Educational Evaluation and Policy Analysis, 34*(4), 435-464.
9. Dee, T. S., & Jacob, B. (2011). The impact of No Child Left Behind on student achievement. *Journal of Policy Analysis and Management, 30*(3), 418-446.
10. Impey, C., Buxner, S., Antonellis, J., Johnson, E., & King, C. (2011). A twenty-year survey of science literacy among college undergraduates. *Journal of College Science Teaching, 40*(4), 31-37.
11. National Science Foundation. (2019). *NSF's Math and Science Partnerships (MSP) program: Strengthening STEM education at all levels.*
12. National Science Foundation. (2020). *Broadening Participation in STEM: A call to action.*

Challenges Faced by Educators and Students

The inadequacies in K-12 science education are not merely the result of systemic issues but also stem from the numerous challenges faced by educators and students in the classroom. These challenges, which range from limited resources and support to societal pressures and misconceptions, can significantly hinder the effectiveness of science instruction and the development of scientific literacy among students.

One of the primary challenges faced by science educators is the lack of adequate resources and support to effectively teach their subject. Many teachers, particularly those in under-resourced schools, struggle with insufficient access to quality instructional materials, such as up-to-date textbooks, laboratory equipment, and technology [1]. This lack of resources can limit teachers' ability to provide hands-on, inquiry-based learning experiences that are essential for fostering deep understanding and engagement with scientific concepts [2].

Moreover, science teachers often face challenges in terms of their own professional development and support. Many teachers, especially at the elementary level, may not have strong backgrounds in science content and pedagogy, which can hinder their ability to deliver effective instruction [3]. Limited opportunities for ongoing professional development, coupled with high levels of teacher turnover and burnout, can further exacerbate this issue [4].

Students, too, face a range of challenges that can impact their engagement and success in science education. One significant challenge is the persistence of stereotypes and misconceptions about science and scientists, which can discourage students, particularly those from underrepresented groups, from pursuing science-related interests and careers [5]. These stereotypes, which often portray scientists as white, male, and socially awkward, can make science feel exclusive and inaccessible to many students [6].

Additionally, students may struggle with the perceived difficulty and abstractness of scientific concepts, leading to frustration and disengagement with the subject [7]. This challenge can be particularly acute for students who have not had sufficient prior exposure to science or who may not have strong foundations in mathematics, which is often essential for understanding and applying scientific principles [8].

The pressure to perform well on standardized tests can also create significant challenges for both educators and students in science classrooms. The emphasis on high-stakes testing can lead to a narrowing of the curriculum, with teachers feeling pressured to "teach to the test" rather than fostering deep understanding and critical thinking skills [9]. This test-driven approach can be particularly detrimental in science education, where hands-on, inquiry-based learning is essential for developing scientific literacy and problem-solving skills [10].

For students from disadvantaged backgrounds, additional challenges such as limited access to resources, parental support, and extracurricular opportunities can further hinder their engagement and success in science education [11]. These students may also face challenges related to language barriers, cultural differences, and the impacts of poverty and trauma on their learning and development [12].

Addressing these challenges will require a multifaceted approach that involves providing educators with the resources, support, and professional development they need to deliver high-quality science instruction. This may involve investments in school infrastructure, instructional materials, and technology, as well as the development of targeted professional development programs that help teachers deepen their content knowledge and pedagogical skills [13].

Efforts to support students in overcoming challenges in science education should focus on creating inclusive, engaging, and relevant learning experiences that help to break down stereotypes and misconceptions about science and scientists. This may involve the use of culturally responsive teaching practices, the integration of real-world applications and problem-solving into the curriculum, and the provision of mentorship and role models from diverse backgrounds [14].

Moreover, addressing the challenges of standardized testing and its impact on science education will require a shift towards more authentic, performance-based assessments that prioritize critical thinking, problem-solving, and the application of scientific knowledge to real-world situations [15]. This may involve the development of new assessment frameworks and the provision of professional development and support for teachers in implementing these assessments effectively.

Ultimately, addressing the challenges faced by educators and students in science education will require a sustained commitment to providing the resources, support, and conditions necessary for effective teaching and learning to take place. By confronting these challenges head-on and working to create more equitable, engaging, and rigorous science education experiences, we can help to foster the development of a more scientifically literate and empowered citizenry.

References

1. National Science Teachers Association. (2013). *The Next Generation Science Standards: For states, by states.* National Academies Press.
2. Whitworth, B. A., & Chiu, J. L. (2015). Professional development and teacher change: The missing leadership link. *Journal of Science Teacher Education, 26*(2), 121-137.
3. Nowicki, B. L., Sullivan-Watts, B., Shim, M. K., Young, B., & Pockalny, R. (2013). Factors influencing science content accuracy in elementary inquiry science lessons. *Research in Science Education, 43*(3), 1135-1154.
4. Ingersoll, R. M., & May, H. (2012). The magnitude, destinations, and determinants of mathematics and science teacher turnover. *Educational Evaluation and Policy Analysis, 34*(4), 435-464.
5. Cheryan, S., Master, A., & Meltzoff, A. N. (2015). Cultural stereotypes as gatekeepers: Increasing girls' interest in computer science and engineering by diversifying stereotypes. *Frontiers in Psychology, 6,* 49.
6. Miller, D. I., Nolla, K. M., Eagly, A. H., & Uttal, D. H. (2018). The development of children's gender-science stereotypes: A meta-analysis of 5 decades of U.S. draw-a-scientist studies. *Child Development, 89*(6), 1943-1955.
7. Osborne, J., Simon, S., & Collins, S. (2003). Attitudes towards science: A review of the literature and its implications. *International Journal of Science Education, 25*(9), 1049-1079.
8. Wang, X. (2013). Why students choose STEM majors: Motivation, high school learning, and postsecondary context of support. *American Educational Research Journal, 50*(5), 1081-1121.
9. Dee, T. S., & Jacob, B. (2011). The impact of No Child Left Behind on student achievement. *Journal of Policy Analysis and Management, 30*(3), 418-446.
10. Penuel, W. R., Harris, C. J., & DeBarger, A. H. (2015). Implementing the Next Generation Science Standards. *Phi Delta Kappan, 96*(6), 45-49.
11. Morgan, P. L., Farkas, G., Hillemeier, M. M., & Maczuga, S. (2016). Science achievement gaps begin very early, persist, and are largely explained by modifiable factors. *Educational Researcher, 45*(1), 18-35.
12. Lee, O., & Buxton, C. A. (2010). *Diversity and equity in science education: Research, policy, and practice.* Teachers College Press.
13. Wilson, S. M. (2013). Professional development for science teachers. *Science, 340*(6130), 310-313.
14. Gay, G. (2010). *Culturally responsive teaching: Theory, research, and practice.* Teachers College Press.
15. Pellegrino, J. W. (2013). Proficiency in science: Assessment challenges and opportunities. *Science, 340*(6130), 320-323.

The Impact of Politicization and Controversies on Science Education

Science education in the United States has long been a subject of political and societal debates, with various controversies and ideological conflicts shaping the content and delivery of science curricula. The politicization of science education has had far-reaching consequences, affecting the quality and effectiveness of science instruction, as well as the public's understanding and trust in scientific knowledge.

One of the most prominent examples of the politicization of science education is the ongoing debate over the teaching of evolution and creationism in public schools. Despite the overwhelming scientific consensus supporting the theory of evolution, many states have faced pressure from religious and conservative groups to include creationism or intelligent design in science curricula as an alternative to evolution [1]. This pressure has led to a number of high-profile legal battles, such as the 1925 Scopes "Monkey" Trial and the 2005 Kitzmiller v. Dover case, which have further intensified the political and social tensions surrounding the teaching of evolution [2].

The politicization of evolution education has had significant impacts on science instruction and student learning. In many cases, teachers may feel pressured to avoid or downplay the teaching of evolution to avoid controversy or backlash from parents or community members [3]. This self-censorship can lead to a watering down of science curricula and a failure to provide students with a comprehensive understanding of one of the foundational theories in biology [4].

Climate change education has also become a highly politicized issue, with some conservative politicians and interest groups seeking to challenge the scientific consensus on human-caused global warming [5]. In recent years, several states have introduced legislation that would allow or require teachers to present "both sides" of the climate change debate, despite the overwhelming scientific evidence supporting the reality and severity of anthropogenic climate change [6].

The politicization of climate change education can create confusion and misconceptions among students, who may be exposed to false or misleading information that contradicts the scientific consensus [7]. Moreover, the framing of climate change as a political controversy rather than a scientific reality can undermine students' understanding of the urgent need for action to mitigate the impacts of global warming [8].

Beyond these specific controversies, the broader politicization of science and science education can have a corrosive effect on public trust in scientific institutions and expertise. When scientific knowledge is treated as a matter of political opinion rather than empirical fact, it becomes easier for misinformation and pseudoscience to gain traction in public discourse [9].

This erosion of trust in science can have serious consequences for public health and well-being, as evidenced by the widespread skepticism and resistance to scientific guidance during the COVID-19 pandemic [10]. When large segments of the population dismiss or reject scientific evidence and recommendations, it becomes difficult to mount an effective societal response to pressing challenges such as infectious disease outbreaks, environmental degradation, and climate change [11].

Addressing the politicization of science education will require a concerted effort from educators, policymakers, and the scientific community to reaffirm the importance of scientific integrity and evidence-based decision-making. This may involve pushing back against efforts to undermine or misrepresent scientific consensus, as well as working to build public trust and understanding of the scientific process [12].

Science educators have a particularly important role to play in this regard, as they are on the front lines of communicating scientific knowledge to students and shaping their attitudes towards science. By presenting science as a dynamic, evidence-based process of inquiry, rather than a set of fixed facts or beliefs, educators can help students develop the critical thinking skills and scientific literacy necessary to navigate complex societal issues [13].

Moreover, science educators can work to create inclusive and culturally responsive learning environments that acknowledge and address the ways in which science has been used to perpetuate social inequities and marginalize certain groups [14]. By confronting the historical and contemporary biases within the scientific community and emphasizing the contributions of diverse scientists, educators can help to break down stereotypes and make science feel more accessible and relevant to all students [15].

Ultimately, depoliticizing science education will require a sustained commitment to upholding the values of scientific integrity, evidence-based decision-making, and open inquiry. By working to build public trust in science and to communicate the importance of scientific literacy for informed citizenship, educators and policymakers can help to create a more resilient and scientifically engaged society.

References

1. Berkman, M., & Plutzer, E. (2010). Evolution, creationism, and the battle to control America's classrooms. Cambridge University Press.
2. Humes, E. (2007). Monkey girl: Evolution, education, religion, and the battle for America's soul. Harper Perennial.
3. Goldston, M. J. D., & Kyzer, P. (2009). Teaching evolution: Narratives with a view from three southern biology teachers in the USA. Journal of Research in Science Teaching, 46(7), 762-790.
4. Glaze, A. L., & Goldston, M. J. (2015). Evolution and science teaching and learning in the United States: A critical review of literature 2000-2013. Science Education, 99(3), 500-518.
5. McCright, A. M., & Dunlap, R. E. (2011). The politicization of climate change and polarization in the American public's views of global warming, 2001–2010. The Sociological Quarterly, 52(2), 155-194.
6. Branch, G., Rosenau, J., & Berbeco, M. (2016). Climate education in the classroom: Cloudy with a chance of confusion. Bulletin of the Atomic Scientists, 72(2), 89-96.
7. Plutzer, E., McCaffrey, M., Hannah, A. L., Rosenau, J., Berbeco, M., & Reid, A. H. (2016). Climate confusion among U.S. teachers. Science, 351(6274), 664-665.
8. Schreiner, C., Henriksen, E. K., & Kirkeby Hansen, P. J. (2005). Climate education: Empowering today's youth to meet tomorrow's challenges. Studies in Science Education, 41(1), 3-49.
9. Hornsey, M. J., Harris, E. A., & Fielding, K. S. (2018). The psychological roots of anti-vaccination attitudes: A 24-nation investigation. Health Psychology, 37(4), 307-315.
10. Rutjens, B. T., van der Linden, S., & van der Lee, R. (2021). Science skepticism in times of COVID-19. Group Processes & Intergroup Relations, 24(2), 276-283.
11. Gauchat, G. (2012). Politicization of science in the public sphere: A study of public trust in the United States, 1974 to 2010. American Sociological Review, 77(2), 167-187.
12. Feinstein, N. W., & Waddington, D. I. (2020). Individual truth judgments or purposeful, collective sensemaking? Rethinking science education's response to the post-truth era. Educational Psychologist, 55(3), 155-166.
13. Osborne, J. (2014). Scientific practices and inquiry in the science classroom. In N. G. Lederman & S. K. Abell (Eds.), Handbook of research on science education (pp. 593-613). Routledge.
14. Bang, M., & Medin, D. (2010). Cultural processes in science education: Supporting the navigation of multiple epistemologies. Science Education, 94(6), 1008-1026.
15. Lee, O., & Buxton, C. A. (2010). Diversity and equity in science education: Research, policy, and practice. Teachers College Press.

Chapter 3: Media, Misinformation, and Pseudoscience

The Role of Media in Shaping Public Understanding of Science

In an era where information is more accessible than ever before, the media plays a crucial role in shaping public understanding and perception of science. From traditional news outlets to social media platforms, the way scientific information is communicated and framed can have a profound impact on how the public engages with and interprets scientific knowledge.

One of the primary functions of the media in relation to science is to act as a conduit between the scientific community and the general public. Journalists and science communicators are tasked with translating complex scientific findings and concepts into language that is accessible and engaging to a broad audience [1]. When done effectively, science journalism can help to increase public interest in and understanding of scientific issues, as well as foster a greater appreciation for the role of science in society [2].

However, the media's coverage of science is not always accurate, comprehensive, or unbiased. In many cases, the pressure to generate attention-grabbing headlines and stories can lead to the oversimplification or sensationalization of scientific findings [3]. This can be particularly problematic when it comes to communicating the inherent uncertainties and nuances of scientific research, which may not fit neatly into a narrative of clear-cut conclusions or breakthroughs [4].

Moreover, the media's selection and framing of scientific stories can be influenced by a range of factors, including editorial biases, commercial interests, and political agendas [5]. This can result in the disproportionate coverage of certain scientific topics or perspectives, while others are overlooked or marginalized. For example, studies have shown that media coverage of climate change has often given undue attention to skeptics and deniers, creating a false sense of balance and uncertainty around the scientific consensus [6].

The rise of digital media and social networking platforms has further complicated the media's role in shaping public understanding of science. While these technologies have democratized access to scientific information and enabled new forms of public engagement with science, they have also created new challenges in terms of misinformation and the spread of pseudoscience [7].

Social media algorithms, which are designed to maximize user engagement and retention, can create echo chambers and filter bubbles that reinforce existing beliefs and biases [8]. This can lead to the rapid spread of false or misleading information about scientific issues, as users are more likely to encounter and share content that aligns with their preexisting views [9]. Moreover, the lack of traditional gatekeepers and quality control mechanisms on social media platforms can make it difficult for users to distinguish between credible scientific information and unsubstantiated claims or conspiracy theories [10].

The impact of misinformation and pseudoscience on public understanding of science can be significant and far-reaching. When large segments of the population are exposed to and believe in false or misleading information about scientific issues, it can undermine public trust in science and make it more difficult to implement evidence-based policies and practices [11]. This has been particularly evident during the COVID-19 pandemic, where the spread of misinformation about the virus, treatments, and vaccines has hindered public health efforts and led to devastating consequences [12].

Addressing the challenges posed by media coverage and misinformation in shaping public understanding of science will require a multifaceted approach. One key strategy is to improve the quality and accuracy of science journalism by providing journalists and communicators with the training, resources, and support they need to effectively cover scientific topics [13]. This may involve collaborations between scientists, journalists, and educators to develop best practices for science communication and to foster a greater understanding of the scientific process among media professionals [14].

Another important approach is to promote media literacy and critical thinking skills among the general public. By equipping individuals with the tools to evaluate the credibility and reliability of scientific information they encounter in the media, we can help to build resilience against misinformation and pseudoscience [15]. This may involve integrating media literacy education into school curricula, as well as developing public awareness campaigns and resources to help people navigate the complex media landscape [16].

Finally, combating the spread of misinformation and pseudoscience on social media platforms will require a concerted effort from tech companies, policymakers, and the scientific community. This may involve developing more robust fact-checking and content moderation systems, as well as collaborating with researchers to better understand and address the psychological and social factors that contribute to the spread of false information [17].

Ultimately, the media's role in shaping public understanding of science is complex and multifaceted, with both opportunities and challenges. By working to improve the quality and accuracy of science journalism, promote media literacy and critical thinking skills, and combat the spread of misinformation and pseudoscience, we can help to foster a more informed and engaged public that is better equipped to make evidence-based decisions and participate in the democratic process.

References

1. Dunwoody, S. (2014). Science journalism: Prospects in the digital age. In M. Bucchi & B. Trench (Eds.), Routledge handbook of public communication of science and technology (pp. 27-39). Routledge.
2. Nisbet, M. C., & Scheufele, D. A. (2009). What's next for science communication? Promising directions and lingering distractions. American Journal of Botany, 96(10), 1767-1778.
3. Weingart, P., & Guenther, L. (2016). Science communication and the issue of trust. Journal of Science Communication, 15(5), C01.
4. Stocking, S. H. (1999). How journalists deal with scientific uncertainty. In S. M. Friedman, S. Dunwoody, & C. L. Rogers (Eds.), Communicating uncertainty: Media coverage of new and controversial science (pp. 23-42). Routledge.
5. Boykoff, M. T., & Boykoff, J. M. (2007). Climate change and journalistic norms: A case-study of US mass-media coverage. Geoforum, 38(6), 1190-1204.
6. Boykoff, M. T. (2013). Public enemy no. 1? Understanding media representations of outlier views on climate change. American Behavioral Scientist, 57(6), 796-817.
7. Brossard, D. (2013). New media landscapes and the science information consumer. Proceedings of the National Academy of Sciences, 110(Supplement 3), 14096-14101.
8. Pariser, E. (2011). The filter bubble: What the Internet is hiding from you. Penguin Press.
9. Vosoughi, S., Roy, D., & Aral, S. (2018). The spread of true and false news online. Science, 359(6380), 1146-1151.
10. Scheufele, D. A., & Krause, N. M. (2019). Science audiences, misinformation, and fake news. Proceedings of the National Academy of Sciences, 116(16), 7662-7669.
11. Lewandowsky, S., Ecker, U. K., & Cook, J. (2017). Beyond misinformation: Understanding and coping with the "post-truth" era. Journal of Applied Research in Memory and Cognition, 6(4), 353-369.
12. Caulfield, T. (2020). Pseudoscience and COVID-19 — we've had enough already. Nature, 580(7802), 182.
13. Makri, A. (2017). Give the public the tools to trust scientists. Nature, 541(7637), 261.
14. Besley, J. C., & Tanner, A. H. (2011). What science communication scholars think about training scientists to communicate. Science Communication, 33(2), 239-263.
15. Vraga, E. K., & Tully, M. (2021). News literacy, social media behaviors, and skepticism toward information on social media. Information, Communication & Society, 24(2), 150-166.
16. Breakstone, J., McGrew, S., Smith, M., Ortega, T., & Wineburg, S. (2018). Teaching students to navigate the online landscape. Social Education, 82(6), 219-221.
17. Lazer, D. M., Baum, M. A., Benkler, Y., Berinsky, A. J., Greenhill, K. M., Menczer, F., ... & Zittrain, J. L. (2018). The science of fake news. Science, 359(6380), 1094-1096.

The Spread of Misinformation and Pseudoscience

In the age of the internet and social media, the spread of misinformation and pseudoscience has become a pervasive and increasingly complex problem. Misinformation, which refers to false or inaccurate information that is spread unintentionally, and disinformation, which is deliberately misleading or biased information, can have serious consequences for public understanding of science and evidence-based decision-making [1].

One of the key factors contributing to the spread of misinformation and pseudoscience is the ease and speed with which information can be shared and amplified online. Social media platforms, in particular, have become powerful tools for disseminating information to large audiences, but they have also created new challenges in terms of quality control and fact-checking [2]. The viral nature of social media means that false or misleading information can spread rapidly and reach millions of people before it can be effectively debunked or corrected [3].

Another factor that contributes to the spread of misinformation and pseudoscience is the phenomenon of confirmation bias, which refers to the tendency of individuals to seek out and interpret information in a way that confirms their preexisting beliefs and attitudes [4]. In the context of science communication, confirmation bias can lead people to selectively engage with information that supports their views while dismissing or ignoring evidence that contradicts them [5]. This can create a self-reinforcing cycle of misinformation, where exposure to false or misleading information reinforces existing beliefs and makes individuals more susceptible to further misinformation [6].

The spread of misinformation and pseudoscience can also be fueled by a lack of scientific literacy and critical thinking skills among the general public. When individuals lack the knowledge and tools to evaluate the credibility and reliability of scientific information, they may be more vulnerable to false or misleading claims [7]. This is particularly problematic in an era where the volume and complexity of scientific information available to the public have increased dramatically, making it more challenging for individuals to navigate and make sense of conflicting or uncertain information [8].

The consequences of misinformation and pseudoscience can be significant and far-reaching. In the context of public health, for example, the spread of false or misleading information about vaccines has contributed to declining vaccination rates and the resurgence of preventable diseases in some communities [9]. Similarly, misinformation about climate change has hindered efforts to address the urgent threat of global warming and build support for evidence-based policies and solutions [10].

Combating the spread of misinformation and pseudoscience requires a multifaceted approach that involves efforts from scientists, communicators, policymakers, and the general public. One key strategy is to improve science communication and public engagement with science. This may involve developing more effective ways of communicating scientific information to diverse audiences, such as using storytelling, visual aids, and interactive formats to make complex concepts more accessible and engaging [11]. It may also involve building stronger partnerships between scientists and journalists, as well as fostering greater public participation in scientific research and decision-making processes [12].

Another important approach is to promote scientific literacy and critical thinking skills among the general public. This may involve integrating science education and media literacy training into school curricula, as well as developing public awareness campaigns and resources to help individuals evaluate the credibility and reliability of scientific information they encounter online [13]. By equipping people with the tools to think critically about scientific claims and evidence, we can help to build resilience against misinformation and pseudoscience [14].

Addressing the spread of misinformation and pseudoscience on social media platforms will also require collaboration between researchers, tech companies, and policymakers. This may involve developing more sophisticated algorithms and content moderation systems to identify and flag false or misleading information, as well as partnering with fact-checking organizations and scientific experts to provide accurate and reliable information to users [15]. It may also involve exploring new approaches to platform design and governance that prioritize transparency, accountability, and the promotion of high-quality information [16].

Ultimately, combating the spread of misinformation and pseudoscience will require a sustained and coordinated effort from all sectors of society. By improving science communication, promoting scientific literacy and critical thinking skills, and working to build a more resilient and reliable information ecosystem, we can help to create a more informed and engaged public that is better equipped to navigate the complex challenges of the 21st century.

References

1. Lazer, D. M., Baum, M. A., Benkler, Y., Berinsky, A. J., Greenhill, K. M., Menczer, F., ... & Zittrain, J. L. (2018). The science of fake news. Science, 359(6380), 1094-1096.
2. Vosoughi, S., Roy, D., & Aral, S. (2018). The spread of true and false news online. Science, 359(6380), 1146-1151.
3. Sharma, M., Yadav, K., Yadav, N., & Ferdinand, K. C. (2017). Zika virus pandemic—analysis of Facebook as a social media health information platform. American Journal of Infection Control, 45(3), 301-302.
4. Nickerson, R. S. (1998). Confirmation bias: A ubiquitous phenomenon in many guises. Review of General Psychology, 2(2), 175-220.
5. Scheufele, D. A., & Krause, N. M. (2019). Science audiences, misinformation, and fake news. Proceedings of the National Academy of Sciences, 116(16), 7662-7669.
6. Del Vicario, M., Bessi, A., Zollo, F., Petroni, F., Scala, A., Caldarelli, G., ... & Quattrociocchi, W. (2016). The spreading of misinformation online. Proceedings of the National Academy of Sciences, 113(3), 554-559.
7. Sharon, A. J., & Baram-Tsabari, A. (2020). Can science literacy help individuals identify misinformation in everyday life?. Science Education, 104(5), 873-894.
8. Breakstone, J., McGrew, S., Smith, M., Ortega, T., & Wineburg, S. (2018). Teaching students to navigate the online landscape. Social Education, 82(4), 219-221.
9. Dubé, E., Vivion, M., & MacDonald, N. E. (2015). Vaccine hesitancy, vaccine refusal and the anti-vaccine movement: influence, impact and implications. Expert Review of Vaccines, 14(1), 99-117.
10. Van der Linden, S., Leiserowitz, A., Rosenthal, S., & Maibach, E. (2017). Inoculating the public against misinformation about climate change. Global Challenges, 1(2), 1600008.
11. Dahlstrom, M. F. (2014). Using narratives and storytelling to communicate science with nonexpert audiences. Proceedings of the National Academy of Sciences, 111(Supplement 4), 13614-13620.
12. Haywood, B. K., & Besley, J. C. (2014). Education, outreach, and inclusive engagement: Towards integrated indicators of successful program outcomes in participatory science. Public Understanding of Science, 23(1), 92-106.
13. Vraga, E. K., & Tully, M. (2021). News literacy, social media behaviors, and skepticism toward information on social media. Information, Communication & Society, 24(2), 150-166.
14. Wineburg, S., McGrew, S., Breakstone, J., & Ortega, T. (2016). Evaluating information: The cornerstone of civic online reasoning. Stanford Digital Repository.
15. Resnick, P., Ovadya, A., & Gilchrist, G. (2019). Iffy Quotient: A platform health metric for misinformation. arXiv preprint arXiv:1902.09982.
16. Lewandowsky, S., Smillie, L., Garcia, D., Hertwig, R., Weatherall, J., Egidy, S., ... & Leiser, M. (2020). Technology and democracy: Understanding the influence of online technologies on political behaviour and decision-making. Publications Office of the European Union.

The Influence of Social Media and the Internet

The advent of the internet and the rise of social media have fundamentally transformed the way people access, share, and engage with information, including scientific knowledge. While these technologies have created unprecedented opportunities for public participation in science and the democratization of knowledge, they have also given rise to new challenges in terms of the spread of misinformation, the proliferation of pseudoscience, and the erosion of trust in scientific expertise [1].

One of the most significant ways in which social media and the internet have influenced the public understanding of science is by drastically changing the information landscape. In the pre-internet era, the dissemination of scientific information was largely controlled by traditional gatekeepers, such as journalists, editors, and scientific institutions, who played a critical role in filtering and curating information for public consumption [2]. However, the rise of digital platforms has disrupted this model, allowing anyone with an internet connection to create, share, and consume information without the same level of oversight or quality control [3].

This democratization of information has had both positive and negative consequences for science communication. On the one hand, it has enabled scientists, science communicators, and members of the public to engage in direct dialogue and share knowledge more easily than ever before [4]. Social media platforms, in particular, have provided new avenues for scientists to communicate their research, build public trust, and foster greater understanding and appreciation of science [5].

However, the open and decentralized nature of the internet has also created a fertile ground for the spread of misinformation and pseudoscience. Without the traditional filters and fact-checking mechanisms, false or misleading information can spread rapidly and unchecked, reaching vast audiences before it can be effectively debunked or corrected [6]. This is particularly problematic in the context of science, where complex and nuanced information can be easily oversimplified, sensationalized, or taken out of context [7].

The influence of social media algorithms and personalization techniques has further compounded this problem. Many social media platforms use algorithms that are designed to maximize user engagement and retention by showing users content that is likely to capture their attention and align with their preexisting beliefs and preferences [8]. This can create echo chambers and filter bubbles, where users are exposed primarily to information that confirms their existing views while being sheltered from opposing perspectives or contradictory evidence [9].

The consequences of this algorithmic filtering can be particularly damaging in the context of science, where exposure to diverse perspectives and evidence is crucial for informed decision-making and the advancement of knowledge [10]. When users are confined to information silos that reinforce their preexisting beliefs, they may become more susceptible to misinformation and pseudoscience, and less likely to engage with credible scientific information that challenges their views [11].

Another way in which social media and the internet have influenced the public understanding of science is by changing the way people evaluate and trust scientific information. In the past, the credibility of scientific information was largely determined by the reputation and authority of the source, such as peer-reviewed journals or respected scientific institutions [12]. However, in the digital age, the proliferation of information sources and the ease of self-publishing have made it more difficult for individuals to assess the credibility and reliability of scientific claims [13].

This has led to a phenomenon known as "source ambiguity," where the origin and credibility of information become unclear or confused [14]. In the context of social media, where information is often shared and re-shared without clear attribution or context, it can be challenging for users to determine the original source or assess the expertise and trustworthiness of the individuals or organizations making scientific claims [15].

Moreover, the social and emotional dynamics of online communication can further complicate the evaluation of scientific information. Studies have shown that people are more likely to trust and share information that is socially validated by their peers, even if that information is false or misleading [16]. The desire for social acceptance and the fear of social rejection can lead individuals to conform to the beliefs and attitudes of their online communities, regardless of the scientific merit of those beliefs [17].

To address the challenges posed by social media and the internet for science communication, it is essential to develop new strategies and approaches that prioritize the promotion of scientific literacy, critical thinking, and information evaluation skills. This may involve partnerships between scientists, science communicators, and digital platforms to create new tools and resources for fact-checking and source verification, as well as efforts to improve the algorithmic curation and recommendation of scientific content [18].

It may also involve greater public engagement and dialogue between scientists and the communities they serve, to build trust, understanding, and a shared sense of ownership over scientific knowledge [19]. By fostering a culture of openness, transparency, and mutual respect, scientists and science communicators can work to counteract the polarizing and distorting effects of online misinformation and pseudoscience [20].

Ultimately, navigating the complex and rapidly evolving landscape of social media and the internet will require ongoing collaboration, experimentation, and adaptation from all stakeholders involved in science communication. By working together to harness the power of these technologies while mitigating their risks and limitations, we can create a more informed, engaged, and scientifically literate society.

References

1. Brossard, D. (2013). New media landscapes and the science information consumer. Proceedings of the National Academy of Sciences, 110(Supplement 3), 14096-14101.
2. Lewandowsky, S., Ecker, U. K., & Cook, J. (2017). Beyond misinformation: Understanding and coping with the "post-truth" era. Journal of Applied Research in Memory and Cognition, 6(4), 353-369.
3. Del Vicario, M., Bessi, A., Zollo, F., Petroni, F., Scala, A., Caldarelli, G., ... & Quattrociocchi, W. (2016). The spreading of misinformation online. Proceedings of the National Academy of Sciences, 113(3), 554-559.
4. Ke, Q., Ahn, Y. Y., & Sugimoto, C. R. (2017). A systematic identification and analysis of scientists on Twitter. PLoS One, 12(4), e0175368.
5. Collins, K., Shiffman, D., & Rock, J. (2016). How are scientists using social media in the workplace?. PLoS One, 11(10), e0162680.
6. Vosoughi, S., Roy, D., & Aral, S. (2018). The spread of true and false news online. Science, 359(6380), 1146-1151.
7. Scheufele, D. A., & Krause, N. M. (2019). Science audiences, misinformation, and fake news. Proceedings of the National Academy of Sciences, 116(16), 7662-7669.
8. Pariser, E. (2011). The filter bubble: What the Internet is hiding from you. Penguin Press.
9. Bakshy, E., Messing, S., & Adamic, L. A. (2015). Exposure to ideologically diverse news and opinion on Facebook. Science, 348(6239), 1130-1132.
10. Stroud, N. J. (2010). Polarization and partisan selective exposure. Journal of Communication, 60(3), 556-576.
11. Schmidt, A. L., Zollo, F., Scala, A., Betsch, C., & Quattrociocchi, W. (2018). Polarization of the vaccination debate on Facebook. Vaccine, 36(25), 3606-3612.
12. Metzger, M. J., & Flanagin, A. J. (2013). Credibility and trust of information in online environments: The use of cognitive heuristics. Journal of Pragmatics, 59, 210-220.
13. Eysenbach, G. (2008). Credibility of health information and digital media: New perspectives and implications for youth. In M. J. Metzger & A. J. Flanagin (Eds.), Digital media, youth, and credibility (pp. 123-154). MacArthur Foundation Series on Digital Media and Learning. MIT Press.
14. Vraga, E. K., & Bode, L. (2017). Using expert sources to correct health misinformation in social media. Science Communication, 39(5), 621-645.
15. Flanagin, A. J., & Metzger, M. J. (2007). The role of site features, user attributes, and information verification behaviors on the perceived credibility of web-based information. New Media & Society, 9(2), 319-342.
16. Lazer, D. M., Baum, M. A., Benkler, Y., Berinsky, A. J., Greenhill, K. M., Menczer, F., ... & Zittrain, J. L. (2018). The science of fake news. Science, 359(6380), 1094-1096.
17. Del Vicario, M., Scala, A., Caldarelli, G., Stanley, H. E., & Quattrociocchi, W. (2017). Modeling confirmation bias and polarization. Scientific Reports, 7, 40391.
18. Wineburg, S., McGrew, S., Breakstone, J., & Ortega, T. (2016). Evaluating information: The cornerstone of civic online reasoning. Stanford Digital Repository.
19. Haywood, B. K., & Besley, J. C. (2014). Education, outreach, and inclusive engagement: Towards integrated indicators of successful program outcomes in participatory science. Public Understanding of Science, 23(1), 92-106.
20. Scheufele, D. A. (2013). Communicating science in social settings. Proceedings of the National Academy of Sciences, 110(Supplement 3), 14040-14047.

Part II:
The Consequences of Scientific Illiteracy

Chapter 4: Public Health and Personal Well-being

Misconceptions about Diseases, Vaccines, and Treatments

The lack of scientific literacy among the general public has significant implications for personal and public health. Misconceptions about diseases, vaccines, and treatments can lead to poor health choices, delayed medical care, and the spread of preventable illnesses [1]. In an era where information is readily available but not always reliable, it is crucial to address these misconceptions and promote evidence-based health practices.

One of the most pervasive misconceptions in public health is the belief that vaccines cause autism. This notion originated from a now-discredited and retracted study published in 1998 by Andrew Wakefield, which suggested a link between the measles, mumps, and rubella (MMR) vaccine and autism [2]. Despite numerous subsequent studies debunking this claim and demonstrating the safety and effectiveness of vaccines, the misconception persists [3].

The consequences of this misconception can be severe. Parents who refuse to vaccinate their children based on unfounded fears put not only their own children at risk but also contribute to the erosion of herd immunity, which protects vulnerable populations who cannot be vaccinated due to age or medical conditions [4]. Vaccine hesitancy has led to resurgences of preventable diseases, such as measles outbreaks in various parts of the world, leading to unnecessary illness, disability, and even death [5].

Another common misconception revolves around the use of antibiotics. Many people believe that antibiotics are effective against viral infections, such as the common cold or flu [6]. This misunderstanding leads to the overuse and misuse of antibiotics, which can contribute to the development of antibiotic-resistant bacteria [7]. When antibiotics are used inappropriately, they can eliminate beneficial bacteria in the body and allow resistant strains to flourish, making infections more difficult to treat [8].

The rise of antimicrobial resistance is a global health threat that is exacerbated by public misconceptions about the appropriate use of antibiotics [9]. Engaging in practices such as taking antibiotics for viral infections, not completing the full course of prescribed antibiotics, or sharing antibiotics with others can all contribute to the problem [10]. Educating the public about the difference between bacterial and viral infections and the proper use of antibiotics is crucial in combating this issue.

Misconceptions about chronic diseases, such as cancer and diabetes, can also have detrimental effects on personal health and well-being. For example, some people believe that cancer is a death sentence and that there is little that can be done to prevent or treat it [11]. This fatalistic view can lead to delays in seeking medical attention, adherence to unproven or dangerous alternative therapies, and a lack of engagement in preventive behaviors, such as regular check-ups and screenings [12].

Similarly, misconceptions about diabetes management, such as the belief that people with diabetes cannot consume any sugar or that insulin causes complications, can lead to poor self-care practices and adverse health outcomes [13]. Providing accurate information about the nature of chronic diseases, risk factors, and evidence-based management strategies is essential for empowering individuals to make informed health decisions and improve their quality of life [14].

The spread of health-related misconceptions can be attributed to various factors, including the proliferation of misinformation online, the influence of celebrity endorsements, and the complexity of scientific information [15]. Social media platforms, in particular, have become a breeding ground for health misinformation, as they allow for the rapid dissemination of unverified claims and anecdotal evidence [16]. The algorithms used by these platforms can create echo chambers that reinforce existing beliefs and make it difficult for users to encounter credible, science-based information [17].

Combating health-related misconceptions requires a multi-pronged approach that involves collaboration among healthcare professionals, public health organizations, media outlets, and educators [18]. Strategies may include:

- Developing and disseminating accessible, engaging, and evidence-based health information through various channels, such as websites, social media, and community outreach programs [19].
- Encouraging healthcare professionals to engage in proactive, patient-centered communication about health topics and to address misconceptions during clinical encounters [20].
- Partnering with educators to integrate health literacy and critical thinking skills into school curricula, equipping students with the tools to evaluate health information and make informed decisions [21].
- Working with media outlets and social media platforms to prioritize the promotion of credible health information and to develop policies and tools to identify and mitigate the spread of misinformation [22].
- Conducting research to better understand the psychological, social, and cultural factors that contribute to the development and persistence of health misconceptions and to develop targeted interventions [23].

By addressing misconceptions about diseases, vaccines, and treatments, we can create a more scientifically literate society that is better equipped to make evidence-based health decisions. Promoting health literacy and critical thinking skills can empower individuals to take control of their health and well-being while also contributing to the overall health and resilience of communities.

References

1. Larson, H. J. (2018). The biggest pandemic risk? Viral misinformation. Nature, 562(7727), 309-310.
2. Wakefield, A. J., Murch, S. H., Anthony, A., Linnell, J., Casson, D. M., Malik, M., ... & Walker-Smith, J. A. (1998). Ileal-lymphoid-nodular hyperplasia, non-specific colitis, and pervasive developmental disorder in children. The Lancet, 351(9103), 637-641. (Retracted)
3. Taylor, L. E., Swerdfeger, A. L., & Eslick, G. D. (2014). Vaccines are not associated with autism: an evidence-based meta-analysis of case-control and cohort studies. Vaccine, 32(29), 3623-3629.
4. Fine, P., Eames, K., & Heymann, D. L. (2011). "Herd immunity": a rough guide. Clinical Infectious Diseases, 52(7), 911-916.
5. Phadke, V. K., Bednarczyk, R. A., Salmon, D. A., & Omer, S. B. (2016). Association between vaccine refusal and vaccine-preventable diseases in the United States: a review of measles and pertussis. JAMA, 315(11), 1149-1158.
6. McNulty, C. A., Boyle, P., Nichols, T., Clappison, P., & Davey, P. (2007). The public's attitudes to and compliance with antibiotics. Journal of Antimicrobial Chemotherapy, 60(suppl_1), i63-i68.
7. Ventola, C. L. (2015). The antibiotic resistance crisis: part 1: causes and threats. Pharmacy and Therapeutics, 40(4), 277.
8. Blaser, M. J. (2016). Antibiotic use and its consequences for the normal microbiome. Science, 352(6285), 544-545.
9. World Health Organization. (2019). Ten threats to global health in 2019.
10. Lee, C. R., Lee, J. H., Kang, L. W., Jeong, B. C., & Lee, S. H. (2015). Educational effectiveness, target, and content for prudent antibiotic use. BioMed Research International, 2015.
11. Robb, K., Stubbings, S., Ramirez, A., Macleod, U., Austoker, J., Waller, J., ... & Wardle, J. (2009). Public awareness of cancer in Britain: a population-based survey of adults. British Journal of Cancer, 101(2), S18-S23.
12. Shahab, L., McGowan, J. A., Waller, J., & Smith, S. G. (2018). Prevalence of beliefs about actual and mythical causes of cancer and their association with socio-demographic and health-related characteristics: Findings from a cross-sectional survey in England. European Journal of Cancer, 103, 308-316.
13. Mc Sharry, J., Dinneen, S. F., Humphreys, M., & O'Donnell, M. (2019). Barriers and facilitators to attendance at Type 2 diabetes structured education programmes: a qualitative study of educators and attendees. Diabetic Medicine, 36(1), 70-79.
14. Powers, M. A., Bardsley, J., Cypress, M., Duker, P., Funnell, M. M., Fischl, A. H., ... & Vivian, E. (2017). Diabetes self-management education and support in type 2 diabetes: a joint position statement of the American Diabetes Association, the American Association of Diabetes Educators, and the Academy of Nutrition and Dietetics. The Diabetes Educator, 43(1), 40-53.
15. Chou, W. Y. S., Oh, A., & Klein, W. M. (2018). Addressing health-related misinformation on social media. JAMA, 320(23), 2417-2418.
16. Sharma, M., Yadav, K., Yadav, N., & Ferdinand, K. C. (2017). Zika virus pandemic—analysis of Facebook as a social media health information platform. American Journal of Infection Control, 45(3), 301-302.
17. Waszak, P. M., Kasprzycka-Waszak, W., & Kubanek, A. (2018). The spread of medical fake news in social media–the pilot quantitative study. Health Policy and Technology, 7(2), 115-118.
18. Ratzan, S. C., Sommarivac, S., & Rauh, L. (2020). Enhancing global health communication during a crisis: lessons from the COVID-19 pandemic. Public Health Research & Practice, 30(2), 3022010.
19. Eichler, K., Wieser, S., & Brügger, U. (2009). The costs of limited health literacy: a systematic review. International Journal of Public Health, 54(5), 313-324.
20. Coleman, C. A., Hudson, S., & Maine, L. L. (2013). Health literacy practices and educational competencies for health professionals: a consensus study. Journal of Health Communication, 18(sup1), 82-102.

21. Nutbeam, D. (2000). Health literacy as a public health goal: a challenge for contemporary health education and communication strategies into the 21st century. Health Promotion International, 15(3), 259-267.
22. Vraga, E. K., & Bode, L. (2017). Using expert sources to correct health misinformation in social media. Science Communication, 39(5), 621-645.
23. Nyhan, B., & Reifler, J. (2015). Does correcting myths about the flu vaccine work? An experimental evaluation of the effects of corrective information. Vaccine, 33(3), 459-464.

The Impact on Personal Health Decisions and Public Health Policies

The consequences of scientific illiteracy and misconceptions about health extend beyond individual well-being, affecting both personal health decisions and the formation of public health policies. When people lack the knowledge and critical thinking skills necessary to evaluate health information accurately, they are more likely to make poorly informed choices that can have serious implications for their health and the health of their communities [1].

At the individual level, scientific illiteracy can lead to a variety of problematic health behaviors. For example, people who do not understand the principles of nutrition and the importance of a balanced diet may make poor dietary choices, increasing their risk of chronic diseases such as obesity, diabetes, and cardiovascular disease [2]. Similarly, those who are unaware of the dangers of smoking or the importance of regular exercise may be less likely to engage in health-promoting behaviors, putting themselves at greater risk for preventable illnesses [3].

Misconceptions about specific health issues can also have a profound impact on personal health decisions. In the case of cancer, for instance, a lack of understanding about the importance of early detection and the effectiveness of modern treatments may lead some individuals to delay seeking medical attention or to pursue unproven or dangerous alternative therapies [4]. This can result in poorer health outcomes and reduced chances of survival [5].

The impact of scientific illiteracy on personal health decisions is particularly concerning in the context of infectious diseases. When people do not understand how diseases spread or the importance of preventive measures such as vaccination and hand hygiene, they may inadvertently contribute to the transmission of illnesses within their communities [6]. This has been tragically exemplified by the resurgence of measles in many parts of the world, largely due to vaccine hesitancy and misinformation [7].

Beyond individual health choices, scientific illiteracy and misconceptions can also have far-reaching effects on public health policies. When a significant portion of the population holds inaccurate beliefs about health issues, it can create pressure on policymakers to make decisions that are not based on scientific evidence [8]. This can lead to the implementation of ineffective or even harmful policies, as well as the misallocation of resources away from evidence-based interventions [9].

One prominent example of this phenomenon is the ongoing debate surrounding climate change and its impact on public health. Despite the overwhelming scientific consensus that human activities are contributing to global warming and that this poses significant risks to human health, many policymakers and members of the public continue to deny or downplay the reality of climate change [10]. This has hindered efforts to implement policies aimed at mitigating the health impacts of climate change, such as reducing greenhouse gas emissions and strengthening public health infrastructure [11].

The influence of scientific illiteracy on public health policies can also be seen in the realm of drug policy. In many countries, the criminalization of drug use and the "war on drugs" approach have been driven more by moral panic and political expediency than by scientific evidence about what works to reduce drug-related harms [12]. This has led to policies that have had devastating consequences for public health, such as the mass incarceration of drug users, the spread of infectious diseases through unsafe injection practices, and the erosion of trust between communities and public health authorities [13].

To address the impact of scientific illiteracy on personal health decisions and public health policies, it is essential to prioritize efforts to improve health literacy and science communication. This may involve developing targeted educational interventions for different population groups, such as school-based programs for children and adolescents, workplace wellness initiatives for adults, and community outreach programs for marginalized communities [14].

Health professionals also have a critical role to play in promoting scientific literacy and countering misconceptions. By taking the time to explain health information clearly and patiently, and by encouraging patients to ask questions and participate in shared decision-making, healthcare providers can help to empower individuals to make more informed choices about their health [15]. Additionally, health professionals can serve as trusted voices in their communities, helping to disseminate accurate health information and combat misinformation [16].

At the policy level, efforts to improve scientific literacy and evidence-based decision-making are also crucial. This may involve strengthening the role of scientific advisory bodies in the policy-making process, increasing funding for health research and science education, and promoting greater transparency and accountability in the use of scientific evidence in policy decisions [17]. Policymakers can also work to build partnerships with researchers, health professionals, and community organizations to ensure that policies are informed by the best available evidence and are responsive to the needs and concerns of diverse populations [18].

Ultimately, addressing the impact of scientific illiteracy on personal health decisions and public health policies will require a sustained and collaborative effort from individuals, communities, health professionals, and policymakers alike. By working together to promote scientific literacy, critical thinking, and evidence-based decision-making, we can create a healthier, more resilient society that is better equipped to face the complex health challenges of the 21st century.

References

1. Nutbeam, D. (2000). Health literacy as a public health goal: a challenge for contemporary health education and communication strategies into the 21st century. Health Promotion International, 15(3), 259-267.
2. Spronk, I., Kullen, C., Burdon, C., & O'Connor, H. (2014). Relationship between nutrition knowledge and dietary intake. British Journal of Nutrition, 111(10), 1713-1726.
3. Berkman, N. D., Sheridan, S. L., Donahue, K. E., Halpern, D. J., & Crotty, K. (2011). Low health literacy and health outcomes: an updated systematic review. Annals of Internal Medicine, 155(2), 97-107.
4. Shahab, L., McGowan, J. A., Waller, J., & Smith, S. G. (2018). Prevalence of beliefs about actual and mythical causes of cancer and their association with socio-demographic and health-related characteristics: Findings from a cross-sectional survey in England. European Journal of Cancer, 103, 308-316.
5. Robb, K., Stubbings, S., Ramirez, A., Macleod, U., Austoker, J., Waller, J., ... & Wardle, J. (2009). Public awareness of cancer in Britain: a population-based survey of adults. British Journal of Cancer, 101(2), S18-S23.
6. Castro-Sánchez, E., Chang, P. W., Vila-Candel, R., Escobedo, A. A., & Holmes, A. H. (2016). Health literacy and infectious diseases: why does it matter? International Journal of Infectious Diseases, 43, 103-110.
7. Phadke, V. K., Bednarczyk, R. A., Salmon, D. A., & Omer, S. B. (2016). Association between vaccine refusal and vaccine-preventable diseases in the United States: a review of measles and pertussis. JAMA, 315(11), 1149-1158.
8. Oliver, K., Lorenc, T., & Innvær, S. (2014). New directions in evidence-based policy research: a critical analysis of the literature. Health Research Policy and Systems, 12(1), 1-11.
9. Brownson, R. C., Chriqui, J. F., & Stamatakis, K. A. (2009). Understanding evidence-based public health policy. American Journal of Public Health, 99(9), 1576-1583.
10. Watts, N., Amann, M., Arnell, N., Ayeb-Karlsson, S., Beagley, J., Belesova, K., ... & Campbell-Lendrum, D. (2021). The 2020 report of The Lancet Countdown on health and climate change: responding to converging crises. The Lancet, 397(10269), 129-170.
11. Watts, N., Amann, M., Arnell, N., Ayeb-Karlsson, S., Belesova, K., Boykoff, M., ... & Montgomery, H. (2019). The 2019 report of The Lancet Countdown on health and climate change: ensuring that the health of a child born today is not defined by a changing climate. The Lancet, 394(10211), 1836-1878.
12. Wood, E., Werb, D., Kazatchkine, M., Kerr, T., Hankins, C., Gorna, R., ... & Montaner, J. (2010). Vienna Declaration: a call for evidence-based drug policies. The Lancet, 376(9738), 310-312.
13. Csete, J., Kamarulzaman, A., Kazatchkine, M., Altice, F., Balicki, M., Buxton, J., ... & Beyrer, C. (2016). Public health and international drug policy. The Lancet, 387(10026), 1427-1480.
14. Rowlands, G., Shaw, A., Jaswal, S., Smith, S., & Harpham, T. (2017). Health literacy and the social determinants of health: a qualitative model from adult learners. Health Promotion International, 32(1), 130-138.
15. Sørensen, K., Van den Broucke, S., Fullam, J., Doyle, G., Pelikan, J., Slonska, Z., & Brand, H. (2012). Health literacy and public health: a systematic review and integration of definitions and models. BMC Public Health, 12(1), 1-13.
16. Ratzan, S. C., Sommarivac, S., & Rauh, L. (2020). Enhancing global health communication during a crisis: lessons from the COVID-19 pandemic. Public Health Research & Practice, 30(2), 3022010.
17. Oliver, K., Innvar, S., Lorenc, T., Woodman, J., & Thomas, J. (2014). A systematic review of barriers to and facilitators of the use of evidence by policymakers. BMC Health Services Research, 14(1), 1-12.
18. Cairney, P., & Oliver, K. (2017). Evidence-based policymaking is not like evidence-based medicine, so how far should you go to bridge the divide between evidence and policy? Health Research Policy and Systems, 15(1), 1-11.

Case Studies: The Anti-Vaccination Movement and the COVID-19 Response

The real-world implications of scientific illiteracy and misconceptions on public health can be vividly illustrated through case studies of the anti-vaccination movement and the global response to the COVID-19 pandemic. These examples demonstrate how a lack of understanding and trust in science can lead to harmful personal health decisions and undermine efforts to address public health crises.

The anti-vaccination movement has its roots in the now-discredited work of Andrew Wakefield, who in 1998 published a fraudulent study claiming a link between the measles, mumps, and rubella (MMR) vaccine and autism [1]. Despite the study being retracted and numerous subsequent studies debunking the claim, the idea that vaccines are dangerous has persisted and spread, fueled in part by celebrity endorsements, online misinformation, and a broader distrust of scientific and medical authorities [2].

The consequences of the anti-vaccination movement have been severe. As vaccination rates have declined in some communities, outbreaks of preventable diseases such as measles, mumps, and pertussis have resurged [3]. In 2019, the United States reported its highest number of measles cases since 1992, with many of these cases occurring in unvaccinated individuals [4]. These outbreaks not only put the health of unvaccinated individuals at risk but also threaten those who cannot be vaccinated due to age or medical conditions, undermining the herd immunity that protects entire communities [5].

The anti-vaccination movement illustrates how scientific illiteracy and misconceptions can lead people to make health decisions that not only harm themselves but also put others at risk. It also highlights the challenges of countering misinformation once it has taken hold, as efforts by public health authorities and the scientific community to promote the safety and efficacy of vaccines have often been met with skepticism and resistance [6].

The global response to the COVID-19 pandemic provides another case study of the impact of scientific illiteracy and misconceptions on public health. From the early days of the pandemic, the spread of misinformation and conspiracy theories about the virus, its origins, and potential treatments has posed a significant challenge to public health efforts [7].

One particularly harmful misconception has been the promotion of unproven or dangerous treatments for COVID-19, such as hydroxychloroquine and ivermectin. Despite a lack of scientific evidence supporting their efficacy and potential risks associated with their use, these treatments have been touted by influential figures and have gained traction among segments of the public [8]. This has led some individuals to self-medicate with these substances, potentially exposing themselves to serious side effects and delaying seeking appropriate medical care [9].

Misconceptions and conspiracy theories have also fueled resistance to public health measures such as mask-wearing, social distancing, and vaccination. Some individuals have refused to comply with these measures, believing them to be ineffective, unnecessary, or even part of a larger plot to control the population [10]. This resistance has not only put these individuals at greater risk of infection but has also hindered efforts to slow the spread of the virus and protect vulnerable populations [11].

The COVID-19 pandemic has also exposed the consequences of scientific illiteracy and mistrust on a global scale. In some countries, a lack of public understanding and trust in science has contributed to delayed or inadequate government responses, as leaders have downplayed the severity of the virus or promoted unproven treatments [12]. This has led to devastating outcomes, with some countries experiencing high rates of infection and mortality that might have been prevented with a more science-based approach [13].

Addressing the challenges posed by scientific illiteracy and misconceptions in the context of public health crises will require a multifaceted approach. One key strategy is to improve science communication and public engagement. This may involve developing clear, accurate, and accessible messaging about public health issues, working with trusted community leaders and organizations to disseminate information, and proactively addressing misinformation and conspiracy theories as they emerge [14].

Another important approach is to build trust in scientific and medical authorities. This may involve increasing transparency in the scientific process, acknowledging and addressing past failures or breaches of trust, and working to ensure that scientific research and public health policies are guided by evidence and the public good rather than political or commercial interests [15].

Ultimately, the case studies of the anti-vaccination movement and the COVID-19 response underscore the critical importance of scientific literacy and trust in science for protecting and promoting public health. By prioritizing efforts to improve science education, communication, and public engagement, we can work to build a society that is better equipped to make informed health decisions and respond effectively to public health challenges.

References

1. Wakefield, A. J., Murch, S. H., Anthony, A., Linnell, J., Casson, D. M., Malik, M., ... & Walker-Smith, J. A. (1998). RETRACTED: Ileal-lymphoid-nodular hyperplasia, non-specific colitis, and pervasive developmental disorder in children. The Lancet, 351(9103), 637-641.
2. Kata, A. (2012). Anti-vaccine activists, Web 2.0, and the postmodern paradigm–An overview of tactics and tropes used online by the anti-vaccination movement. Vaccine, 30(25), 3778-3789.
3. Phadke, V. K., Bednarczyk, R. A., Salmon, D. A., & Omer, S. B. (2016). Association between vaccine refusal and vaccine-preventable diseases in the United States: a review of measles and pertussis. JAMA, 315(11), 1149-1158.
4. Centers for Disease Control and Prevention. (2020). Measles cases and outbreaks. Retrieved from https://www.cdc.gov/measles/cases-outbreaks.html
5. Fine, P., Eames, K., & Heymann, D. L. (2011). "Herd immunity": a rough guide. Clinical Infectious Diseases, 52(7), 911-916.
6. Larson, H. J., Cooper, L. Z., Eskola, J., Katz, S. L., & Ratzan, S. (2011). Addressing the vaccine confidence gap. The Lancet, 378(9790), 526-535.
7. Zarocostas, J. (2020). How to fight an infodemic. The Lancet, 395(10225), 676.
8. Meyerowitz, E. A., Vannier, A. G., Friesen, M. G., Schoenfeld, S., Gelfand, J. A., Callahan, M. V., ... & Kim, A. Y. (2020). Rethinking the role of hydroxychloroquine in the treatment of COVID-19. The FASEB Journal, 34(5), 6027-6037.
9. Caly, L., Druce, J. D., Catton, M. G., Jans, D. A., & Wagstaff, K. M. (2020). The FDA-approved drug ivermectin inhibits the replication of SARS-CoV-2 in vitro. Antiviral Research, 178, 104787.
10. Romer, D., & Jamieson, K. H. (2020). Conspiracy theories as barriers to controlling the spread of COVID-19 in the US. Social Science & Medicine, 263, 113356.

11. Allington, D., Duffy, B., Wessely, S., Dhavan, N., & Rubin, J. (2020). Health-protective behaviour, social media usage and conspiracy belief during the COVID-19 public health emergency. Psychological Medicine, 1-7.
12. Bump, J. B., Baum, F., Sakornsin, M., Yates, R., & Hofman, K. (2021). Political economy of covid-19: extractive, regressive, competitive. BMJ, 372.
13. Horton, R. (2021). Offline: The lessons of smallpox eradication for COVID-19. The Lancet, 397(10276), 809.
14. Paakkari, L., & Okan, O. (2020). COVID-19: health literacy is an underestimated problem. The Lancet Public Health, 5(5), e249-e250.
15. Larson, H. J. (2020). Blocking information on COVID-19 can fuel the spread of misinformation. Nature, 580(7803), 306.

Chapter 5: Environmental Issues and Climate Change

Misunderstanding of Environmental Science and Climate Change

One of the most pressing and complex challenges facing humanity today is the threat of climate change and environmental degradation. Despite the overwhelming scientific consensus that human activities are driving these problems, there remains a significant gap in public understanding and acceptance of the underlying science. This misunderstanding of environmental science and climate change has far-reaching consequences, impacting individual behaviors, public discourse, and policy decisions.

At the heart of this misunderstanding is a lack of scientific literacy and a prevalence of misconceptions about key environmental concepts. For example, many people struggle to distinguish between weather and climate, leading them to dismiss long-term climate trends based on short-term weather events [1]. This confusion is often exploited by those seeking to sow doubt about the reality and severity of climate change, who point to cold snaps or snowstorms as evidence that global warming is not occurring [2].

Another common misconception is the belief that climate change is a natural phenomenon that has occurred throughout Earth's history, and that current changes are not significantly different or caused by human activities [3]. While it is true that the Earth's climate has varied naturally over long timescales, the current rate and magnitude of change are unprecedented and can be directly attributed to human-caused greenhouse gas emissions [4]. This misconception underlies much of the skepticism and denial

59

of climate change, as well as the argument that human actions are not responsible for the problem.

The misunderstanding of environmental science also extends to other critical issues, such as biodiversity loss, deforestation, and pollution. Many people are unaware of the scale and severity of these problems, or the complex web of ecological interactions that sustain life on Earth [5]. This lack of understanding can lead to a disregard for the importance of conservation and a failure to appreciate the societal and economic benefits of healthy ecosystems [6].

The reasons for this misunderstanding of environmental science are multifaceted and complex. One significant factor is the inherent complexity of ecological systems and the challenges of communicating scientific information to a lay audience. Environmental issues often involve intricate feedbacks, tipping points, and uncertainties that can be difficult to convey in simple terms [7]. Moreover, the timescales over which environmental changes occur can be vast, making it challenging for people to perceive the impacts of their actions or the urgency of the problem [8].

Another factor contributing to the misunderstanding of environmental science is the politicization of the issue and the spread of misinformation. In recent decades, environmental issues, particularly climate change, have become highly polarized along ideological lines [9]. This polarization has been fueled in part by a well-funded and organized campaign of denial and misinformation, often backed by fossil fuel interests and conservative think tanks [10]. The spread of false or misleading information about climate change and other environmental issues has sown confusion and doubt among the public, undermining trust in scientific authorities and hindering efforts to address the problem [11].

The media also plays a significant role in shaping public understanding of environmental science. In the pursuit of balanced reporting, media outlets often give equal weight to scientific evidence and dissenting opinions, even when those opinions are not supported by empirical evidence [12]. This false balance can create the impression that there is significant scientific uncertainty or debate about issues like climate change, when in reality, the consensus among experts is overwhelming [13]. Moreover, the media's tendency to focus on dramatic or sensational aspects of environmental stories, such as natural disasters or doomsday scenarios, can obscure the more gradual and systemic nature of the problem [14].

Overcoming the misunderstanding of environmental science and climate change will require a concerted effort to improve scientific literacy, communication, and public engagement. One key strategy is to prioritize environmental education at all levels, from primary schools to adult learning programs [15]. This education should focus not only on conveying factual knowledge but also on developing critical thinking skills and an understanding of the scientific process [16]. By equipping people with the tools to evaluate scientific information and make informed decisions, we can build a more environmentally literate society.

Another important approach is to improve the communication of environmental science to the public. This may involve developing clear, accessible, and engaging materials that translate complex scientific concepts into language that resonates with different audiences [17]. It may also involve training scientists and environmental professionals in effective communication strategies and encouraging them to engage directly with the public through outreach and community events [18].

Countering misinformation and promoting trust in scientific authorities is also critical. This may involve fact-checking and debunking false claims, as well as proactively communicating the scientific consensus on environmental issues [19]. It may also involve efforts to increase transparency and accountability in environmental research and decision-making, and to build partnerships between scientists, policymakers, and community stakeholders [20].

Ultimately, addressing the misunderstanding of environmental science and climate change will require a fundamental shift in how we communicate and engage with these issues as a society. By prioritizing scientific literacy, evidence-based decision-making, and public participation, we can work towards a more informed and engaged populace that is better equipped to confront the environmental challenges of the 21st century.

References

1. Weber, E. U., & Stern, P. C. (2011). Public understanding of climate change in the United States. American Psychologist, 66(4), 315-328.
2. Bohr, J. (2014). Public views on the dangers and importance of climate change: predicting climate change beliefs in the United States through income moderated by party identification. Climatic Change, 126(1-2), 217-227.
3. Hornsey, M. J., Harris, E. A., Bain, P. G., & Fielding, K. S. (2016). Meta-analyses of the determinants and outcomes of belief in climate change. Nature Climate Change, 6(6), 622-626.
4. Intergovernmental Panel on Climate Change. (2014). Climate change 2014: Synthesis report. Contribution of Working Groups I, II and III to the fifth assessment report of the Intergovernmental Panel on Climate Change.
5. Cardinale, B. J., Duffy, J. E., Gonzalez, A., Hooper, D. U., Perrings, C., Venail, P., ... & Naeem, S. (2012). Biodiversity loss and its impact on humanity. Nature, 486(7401), 59-67.
6. Balmford, A., Bruner, A., Cooper, P., Costanza, R., Farber, S., Green, R. E., ... & Turner, R. K. (2002). Economic reasons for conserving wild nature. Science, 297(5583), 950-953.
7. Pidgeon, N., & Fischhoff, B. (2011). The role of social and decision sciences in communicating uncertain climate risks. Nature Climate Change, 1(1), 35-41.
8. Gifford, R. (2011). The dragons of inaction: psychological barriers that limit climate change mitigation and adaptation. American Psychologist, 66(4), 290-302.
9. McCright, A. M., & Dunlap, R. E. (2011). The politicization of climate change and polarization in the American public's views of global warming, 2001–2010. The Sociological Quarterly, 52(2), 155-194.
10. Oreskes, N., & Conway, E. M. (2011). Merchants of doubt: How a handful of scientists obscured the truth on issues from tobacco smoke to global warming. Bloomsbury Publishing USA.
11. Lewandowsky, S., Ecker, U. K., & Cook, J. (2017). Beyond misinformation: Understanding and coping with the "post-truth" era. Journal of Applied Research in Memory and Cognition, 6(4), 353-369.
12. Boykoff, M. T., & Boykoff, J. M. (2004). Balance as bias: global warming and the US prestige press. Global Environmental Change, 14(2), 125-136.
13. Cook, J., Oreskes, N., Doran, P. T., Anderegg, W. R., Verheggen, B., Maibach, E. W., ... & Rice, K. (2016). Consensus on consensus: a synthesis of consensus estimates on human-caused global warming. Environmental Research Letters, 11(4), 048002.
14. Hulme, M. (2009). Why we disagree about climate change: Understanding controversy, inaction and opportunity. Cambridge University Press.
15. UNESCO. (2014). Shaping the future we want: UN Decade of Education for Sustainable Development (2005-2014) final report.
16. Wals, A. E., Brody, M., Dillon, J., & Stevenson, R. B. (2014). Convergence between science and environmental education. Science, 344(6184), 583-584.
17. Corner, A., Shaw, C., & Clarke, J. (2018). Principles for effective communication and public engagement on climate change: A Handbook for IPCC authors. Climate Outreach.
18. Ramirez-Andreotta, M. D., Brusseau, M. L., Artiola, J. F., Maier, R. M., & Gandolfi, A. J. (2014). Environmental research translation: Enhancing interactions with communities at contaminated sites. Science of the Total Environment, 497, 651-664.
19. Cook, J., Ellerton, P., & Kinkead, D. (2018). Deconstructing climate misinformation to identify reasoning errors. Environmental Research Letters, 13(2), 024018.
20. Dietz, T. (2013). Bringing values and deliberation to science communication. Proceedings of the National Academy of Sciences, 110(Supplement 3), 14081-14087.

The Impact on Individual Actions and Support for Environmental Policies

The misunderstanding of environmental science and climate change has far-reaching consequences that extend beyond the realm of public discourse and into the daily lives and decision-making processes of individuals. When people lack a clear understanding of the causes, impacts, and solutions to environmental problems, they may be less likely to take meaningful action to address these issues or to support policies aimed at mitigating their effects.

At the individual level, a lack of environmental literacy can lead to a range of problematic behaviors and choices. For example, people who are unaware of the environmental impacts of their consumption habits may continue to engage in unsustainable practices, such as using single-use plastics, wasting food, or purchasing products with high carbon footprints [1]. Similarly, those who do not understand the importance of energy conservation may fail to adopt simple measures like turning off lights, adjusting thermostats, or using public transportation, all of which can contribute to reducing greenhouse gas emissions [2].

Misunderstandings about the nature and severity of environmental problems can also lead individuals to prioritize other concerns over environmental protection. When people believe that environmental issues are distant, abstract, or unrelated to their daily lives, they may be less motivated to make personal sacrifices or to support policies that address these problems [3]. This is particularly true when environmental policies are perceived as conflicting with other values or priorities, such as economic growth, personal freedom, or social justice [4].

The impact of environmental illiteracy on individual actions is compounded by the prevalence of psychological barriers that can hinder pro-environmental behavior change. These barriers include factors such as limited cognition, ideological worldviews, comparisons with key other people, sunk costs, discredence toward experts and authorities, perceived risks of change, and positive but inadequate behavior change [5]. Even when individuals are aware of environmental problems and motivated to act, these barriers can lead to inaction, delay, or insufficient responses.

The consequences of individual inaction and lack of support for environmental policies are significant and far-reaching. When large segments of the population continue to engage in unsustainable behaviors and oppose or remain indifferent to environmental policies, it becomes increasingly difficult to address urgent problems like climate change, biodiversity loss, and pollution. This collective inaction can lead to a sense of helplessness or resignation, further reinforcing the belief that individual actions do not matter in the face of such vast and complex challenges [6].

Moreover, the lack of public support for environmental policies can create political obstacles to the implementation of effective solutions. When policymakers perceive that there is insufficient public demand or political will for environmental action, they may be less likely to prioritize these issues or to take bold steps to address them [7]. This can result in a vicious cycle, where the lack of public understanding and engagement leads to weak or ineffective policies, which in turn fail to generate the necessary public support and momentum for more ambitious action.

Overcoming these challenges will require a concerted effort to improve environmental literacy and to promote individual and collective action. One key strategy is to develop more effective ways of communicating environmental science to the public, using language, frames, and narratives that resonate with different audiences and that highlight the personal and societal benefits of environmental protection [8]. This may involve emphasizing the health, economic, and security implications of environmental problems, as well as the opportunities for innovation, job creation, and community resilience that can arise from addressing these challenges [9].

Another important approach is to provide individuals with the tools, resources, and incentives they need to adopt more sustainable behaviors and to support environmental policies. This may include measures such as eco-labeling, green procurement, and subsidies for environmentally friendly products and services, as well as education and outreach programs that help individuals understand the impacts of their choices and the options available to them [10]. It may also involve creating social norms and peer support networks that encourage and reinforce pro-environmental behaviors, and that help individuals overcome psychological barriers to change [11].

Engaging individuals as active participants in environmental decision-making and policy development is also critical. When people feel that they have a stake in the outcomes of environmental policies and that their voices and concerns are being heard, they are more likely to support and comply with these policies [12]. Participatory approaches, such as citizen science, community-based monitoring, and deliberative forums, can help to build trust, understanding, and ownership of environmental issues, and to generate more robust and socially acceptable solutions [13].

Ultimately, addressing the misunderstanding of environmental science and its impact on individual actions and policy support will require a fundamental shift in how we communicate, educate, and engage the public on these issues. By fostering a more environmentally literate and empowered citizenry, we can build the social and political will necessary to tackle the urgent environmental challenges of our time and to create a more sustainable and resilient future for all.

References

1. Ivanova, D., Stadler, K., Steen-Olsen, K., Wood, R., Vita, G., Tukker, A., & Hertwich, E. G. (2016). Environmental impact assessment of household consumption. Journal of Industrial Ecology, 20(3), 526-536.
2. Attari, S. Z., DeKay, M. L., Davidson, C. I., & De Bruin, W. B. (2010). Public perceptions of energy consumption and savings. Proceedings of the National Academy of Sciences, 107(37), 16054-16059.
3. Leiserowitz, A. (2006). Climate change risk perception and policy preferences: The role of affect, imagery, and values. Climatic Change, 77(1-2), 45-72.
4. Drews, S., & Van den Bergh, J. C. (2016). What explains public support for climate policies? A review of empirical and experimental studies. Climate Policy, 16(7), 855-876.
5. Gifford, R. (2011). The dragons of inaction: psychological barriers that limit climate change mitigation and adaptation. American Psychologist, 66(4), 290-302.
6. Norgaard, K. M. (2011). Living in denial: Climate change, emotions, and everyday life. MIT Press.
7. Ockwell, D., Whitmarsh, L., & O'Neill, S. (2009). Reorienting climate change communication for effective mitigation: forcing people to be green or fostering grass-roots engagement? Science Communication, 30(3), 305-327.
8. Nisbet, M. C. (2009). Communicating climate change: Why frames matter for public engagement. Environment: Science and Policy for Sustainable Development, 51(2), 12-23.

9. Bain, P. G., Milfont, T. L., Kashima, Y., Bilewicz, M., Doron, G., Garðarsdóttir, R. B., ... & Saviolidis, N. M. (2016). Co-benefits of addressing climate change can motivate action around the world. Nature Climate Change, 6(2), 154-157.
10. Stern, P. C. (2011). Contributions of psychology to limiting climate change. American Psychologist, 66(4), 303-314.
11. Abrahamse, W., & Steg, L. (2013). Social influence approaches to encourage resource conservation: A meta-analysis. Global Environmental Change, 23(6), 1773-1785.
12. Dietz, T., & Stern, P. C. (2008). Public participation in environmental assessment and decision making. National Academies Press.
13. Bonney, R., Shirk, J. L., Phillips, T. B., Wiggins, A., Ballard, H. L., Miller-Rushing, A. J., & Parrish, J. K. (2014). Next steps for citizen science. Science, 343(6178), 1436-1437.

Consequences for the Planet and Future Generations

The misunderstanding of environmental science and climate change, coupled with the lack of individual and collective action, has profound consequences that extend far beyond the present moment. The failure to address these issues urgently and effectively threatens the health and stability of the planet and the well-being of future generations.

One of the most severe and far-reaching consequences of environmental inaction is the acceleration of climate change. As greenhouse gas emissions continue to rise, driven by human activities such as fossil fuel combustion, deforestation, and industrial processes, the Earth's average surface temperature is increasing at an unprecedented rate [1]. This warming trend is already having devastating impacts on natural systems and human societies, including rising sea levels, more frequent and intense heatwaves, droughts, and extreme weather events, and the loss of biodiversity [2].

If left unchecked, climate change is projected to cause irreversible damage to the planet's ecosystems and to the life-support systems upon which all species depend. The Intergovernmental Panel on Climate Change (IPCC) warns that a global temperature increase of 1.5°C above pre-industrial levels could lead to the loss of up to 90% of coral reefs, the extinction of many plant and animal species, and the displacement of millions of people due to sea-level rise and other climate-related hazards [3]. At higher levels of warming, the risks become even more severe, potentially triggering tipping points and feedback loops that could lead to runaway climate change and the collapse of entire ecosystems [4].

The consequences of climate change will be felt most acutely by future generations, who will inherit a planet that is less stable, less diverse, and less hospitable than the one we inhabit today. Children born in the 21st century are likely to experience a world that is 4°C warmer by the end of their lives, facing unprecedented risks to their health, safety, and prosperity [5]. They will also bear the burden of adapting to and mitigating the impacts of climate change, which will require massive investments in infrastructure, technology, and social resilience [6].

Beyond climate change, the misunderstanding and mismanagement of environmental issues have other dire consequences for the planet and its inhabitants. The loss of biodiversity, driven by habitat destruction, overexploitation, pollution, and invasive species, is eroding the fabric of life on Earth and undermining the ecosystem services that support human well-being [7]. The World Wildlife Fund estimates that populations of vertebrate species have declined by an average of 60% since 1970, with many iconic species such as elephants, rhinos, and tigers facing extinction in the wild [8].

The degradation of natural habitats not only threatens the survival of individual species but also the functioning of entire ecosystems. Forests, wetlands, and other natural areas provide vital services such as carbon sequestration, water filtration, soil formation, and nutrient cycling, which are essential for the health and resilience of the planet [9]. The loss of these services can have cascading effects on human societies, including reduced food security, increased vulnerability to natural disasters, and the emergence of new diseases [10].

The consequences of environmental degradation are not evenly distributed, with poor and marginalized communities often bearing the brunt of the impacts. Environmental injustice, whereby certain populations are disproportionately exposed to pollution, waste, and other environmental hazards, is a growing concern in many parts of the world [11]. Children, in particular, are highly vulnerable to the health impacts of environmental pollution, including respiratory illnesses, developmental disorders, and cancer [12].

The failure to address environmental issues also has profound economic and social implications for future generations. The costs of inaction on climate change alone are estimated to be in the trillions of dollars, with the potential to slash global GDP by up to 20% by the end of the century [13]. These costs will fall disproportionately on future generations, who will have to contend with the compounding effects of environmental degradation, resource depletion, and social instability [14].

To avoid the most catastrophic consequences of environmental inaction, it is imperative that we take bold and urgent steps to address these challenges. This will require a fundamental shift in our understanding of the relationship between human societies and the natural world, as well as a commitment to evidence-based decision-making and long-term thinking [15].

One key strategy is to prioritize the transition to a low-carbon, sustainable economy that values the conservation and restoration of natural resources. This will involve phasing out fossil fuels, investing in renewable energy and green infrastructure, and adopting circular economy principles that minimize waste and pollution [16]. It will also require protecting and restoring critical ecosystems, such as forests, wetlands, and coral reefs, which provide essential services for climate regulation, biodiversity conservation, and human well-being [17].

Equally important is the need to promote environmental literacy and to engage the public in the co-creation of solutions. By fostering a deeper understanding of the complex interconnections between human activities and ecological systems, we can empower individuals and communities to make informed decisions and to take action in support of a more sustainable future [18]. This will require investing in education, communication, and outreach efforts that reach diverse audiences and that highlight the benefits of environmental stewardship for human health, prosperity, and security [19].

Ultimately, safeguarding the planet and the well-being of future generations will require a collective effort that transcends political, cultural, and geographic boundaries. It will demand a recognition of our shared responsibility as global citizens and a commitment to the principles of intergenerational equity and environmental justice [20]. By rising to this challenge and taking decisive action to address the misunderstanding and mismanagement of environmental issues, we can help to ensure a more livable, equitable, and sustainable world for ourselves and for the generations to come.

References

1. Cook, J., Oreskes, N., Doran, P. T., Anderegg, W. R., Verheggen, B., Maibach, E. W., ... & Rice, K. (2016). Consensus on consensus: a synthesis of consensus estimates on human-caused global warming. Environmental Research Letters, 11(4), 048002.
2. Intergovernmental Panel on Climate Change. (2014). Climate change 2014: Synthesis report. Contribution of Working Groups I, II and III to the fifth assessment report of the Intergovernmental Panel on Climate Change.
3. Intergovernmental Panel on Climate Change. (2018). Global warming of 1.5°C. An IPCC Special Report on the impacts of global warming of 1.5°C above pre-industrial levels and related global greenhouse gas emission pathways, in the context of strengthening the global response to the threat of climate change, sustainable development, and efforts to eradicate poverty.
4. Lenton, T. M., Rockström, J., Gaffney, O., Rahmstorf, S., Richardson, K., Steffen, W., & Schellnhuber, H. J. (2019). Climate tipping points—too risky to bet against. Nature, 575(7784), 592-595.
5. Watts, N., Amann, M., Arnell, N., Ayeb-Karlsson, S., Belesova, K., Boykoff, M., ... & Montgomery, H. (2019). The 2019 report of The Lancet Countdown on health and climate change: ensuring that the health of a child born today is not defined by a changing climate. The Lancet, 394(10211), 1836-1878.
6. Stern, N. (2007). The economics of climate change: The Stern review. Cambridge University Press.
7. Cardinale, B. J., Duffy, J. E., Gonzalez, A., Hooper, D. U., Perrings, C., Venail, P., ... & Naeem, S. (2012). Biodiversity loss and its impact on humanity. Nature, 486(7401), 59-67.
8. World Wildlife Fund. (2018). Living Planet Report 2018: Aiming higher.
9. Costanza, R., De Groot, R., Sutton, P., Van der Ploeg, S., Anderson, S. J., Kubiszewski, I., ... & Turner, R. K. (2014). Changes in the global value of ecosystem services. Global Environmental Change, 26, 152-158.
10. Myers, S. S., Gaffikin, L., Golden, C. D., Ostfeld, R. S., Redford, K. H., Ricketts, T. H., ... & Osofsky, S. A. (2013). Human health impacts of ecosystem alteration. Proceedings of the National Academy of Sciences, 110(47), 18753-18760.
11. Bullard, R. D., Mohai, P., Saha, R., & Wright, B. (2008). Toxic wastes and race at twenty: Why race still matters after all of these years. Environmental Law, 38(2), 371-411.
12. Landrigan, P. J., Fuller, R., Acosta, N. J., Adeyi, O., Arnold, R., Baldé, A. B., ... & Zhong, M. (2018). The Lancet Commission on pollution and health. The Lancet, 391(10119), 462-512.
13. Burke, M., Davis, W. M., & Diffenbaugh, N. S. (2018). Large potential reduction in economic damages under UN mitigation targets. Nature, 557(7706), 549-553.
14. Stern, N. (2008). The economics of climate change. American Economic Review, 98(2), 1-37.

15. Rockström, J., Steffen, W., Noone, K., Persson, Å., Chapin III, F. S., Lambin, E., ... & Foley, J. (2009). Planetary boundaries: exploring the safe operating space for humanity. Ecology and Society, 14(2).
16. Ellen MacArthur Foundation. (2013). Towards the circular economy: Economic and business rationale for an accelerated transition.
17. Watson, J. E., Dudley, N., Segan, D. B., & Hockings, M. (2014). The performance and potential of protected areas. Nature, 515(7525), 67-73.
18. Wals, A. E., Brody, M., Dillon, J., & Stevenson, R. B. (2014). Convergence between science and environmental education. Science, 344(6184), 583-584.
19. Nisbet, M. C., & Scheufele, D. A. (2009). What's next for science communication? Promising directions and lingering distractions. American Journal of Botany, 96(10), 1767-1778.
20. Brown Weiss, E. (1989). In fairness to future generations: International law, common patrimony, and intergenerational equity. Transnational Publishers.

Chapter 6: Technological Advancement and Economic Competitiveness

The Importance of Science and Technology for Innovation and Economic Growth

In today's rapidly evolving global landscape, science and technology have emerged as key drivers of innovation and economic growth. The development and application of new knowledge and technologies have the potential to transform industries, create new markets, and address pressing societal challenges. As such, fostering a strong and vibrant science and technology ecosystem has become a critical priority for nations seeking to maintain their competitive edge and ensure long-term prosperity.

At the heart of this ecosystem lies the process of innovation, which involves the translation of scientific discoveries and technological advancements into new products, services, and processes [1]. Innovation fuels economic growth by creating value, enhancing productivity, and opening up new opportunities for businesses and entrepreneurs [2]. It also plays a crucial role in addressing complex social and environmental issues, such as healthcare, energy, and sustainability, by providing novel solutions and approaches [3].

The importance of science and technology for innovation and economic growth can be seen across various sectors and industries. In the healthcare sector, for example, advances in biomedical research and medical technologies have led to the development of new treatments, diagnostic tools, and preventive strategies [4]. These innovations have not only improved patient outcomes and quality of life but have also generated significant economic benefits, such as reducing healthcare costs, creating jobs, and stimulating the growth of the pharmaceutical and medical device industries [5].

Similarly, in the energy sector, scientific research and technological advancements have been instrumental in driving the transition towards cleaner, more sustainable sources of power. The development of renewable energy technologies, such as solar, wind, and hydro power, has helped to reduce greenhouse gas emissions, enhance energy security, and create new economic opportunities [6]. The growth of the clean energy industry has also had ripple effects throughout the economy, stimulating demand for related products and services, such as energy storage systems, smart grid technologies, and green building materials [7].

The information and communication technology (ICT) sector is another prime example of how science and technology have transformed the economic landscape. The rapid pace of innovation in areas such as artificial intelligence, big data, and the Internet of Things has given rise to entirely new industries and business models [8]. These technologies have also had a profound impact on traditional sectors, such as manufacturing, agriculture, and transportation, by enabling greater automation, efficiency, and connectivity [9].

Beyond specific sectors and industries, science and technology have also played a crucial role in driving overall economic growth and competitiveness at the national level. Countries that invest heavily in research and development (R&D), education, and technological infrastructure tend to experience higher levels of innovation, productivity, and economic output [10]. This is because a strong science and technology base provides the foundation for knowledge creation, skill development, and the commercialization of new ideas and technologies [11].

Empirical evidence supports the link between science, technology, and economic performance. Studies have shown that countries with higher levels of R&D spending and scientific output tend to have faster rates of economic growth, greater levels of productivity, and more competitive economies [12]. Moreover, firms that actively engage in innovation and adopt new technologies tend to outperform their peers in terms of revenue growth, market share, and profitability [13].

However, realizing the full potential of science and technology for innovation and economic growth requires more than just investment in R&D. It also requires a supportive policy environment that encourages entrepreneurship, facilitates technology transfer, and promotes collaboration between academia, industry, and government [14]. This includes policies that protect intellectual property rights, reduce regulatory barriers, and provide incentives for private sector investment in innovation [15].

Additionally, fostering a culture of innovation and entrepreneurship is critical for translating scientific and technological advancements into economic value. This involves promoting risk-taking, creativity, and experimentation, as well as providing access to funding, mentorship, and networking opportunities for entrepreneurs and startups [16]. It also involves building a skilled and adaptable workforce that can keep pace with the changing demands of the knowledge economy [17].

Education and training play a vital role in developing the human capital necessary for driving innovation and economic growth. Investing in science, technology, engineering, and mathematics (STEM) education, from primary school through higher education, is essential for cultivating a pipeline of talent that can contribute to the advancement of science and technology [18]. Moreover, promoting interdisciplinary collaboration and problem-based learning can help to foster the critical thinking, creativity, and communication skills that are essential for innovation [19].

Finally, international cooperation and collaboration are becoming increasingly important in the context of science, technology, and innovation. In an era of global challenges and opportunities, nations that work together to share knowledge, resources, and best practices are better positioned to leverage the benefits of science and technology for their mutual advantage [20]. This includes collaborating on large-scale scientific projects, developing international standards and regulations, and promoting the free flow of ideas and talent across borders.

In conclusion, science and technology are essential engines of innovation and economic growth in the modern world. By driving the creation and application of new knowledge and technologies, they have the potential to transform industries, address societal challenges, and create new opportunities for businesses and individuals alike. However, realizing this potential requires a comprehensive and collaborative approach that involves investment, policy support, entrepreneurship, education, and international cooperation. As nations seek to navigate the complex and rapidly evolving landscape of the 21st century, fostering a strong and vibrant science and technology ecosystem will be critical for ensuring their long-term competitiveness and prosperity.

References

1. OECD. (2018). Oslo Manual 2018: Guidelines for collecting, reporting and using data on innovation. OECD Publishing.
2. Solow, R. M. (1957). Technical change and the aggregate production function. The Review of Economics and Statistics, 39(3), 312-320.
3. Anadon, L. D., Chan, G., Harley, A. G., Matus, K., Moon, S., Murthy, S. L., & Clark, W. C. (2016). Making technological innovation work for sustainable development. Proceedings of the National Academy of Sciences, 113(35), 9682-9690.
4. Moses, H., Matheson, D. H., Cairns-Smith, S., George, B. P., Palisch, C., & Dorsey, E. R. (2015). The anatomy of medical research: US and international comparisons. JAMA, 313(2), 174-189.
5. Deloitte. (2019). 2020 global life sciences outlook. Deloitte Insights.
6. IRENA. (2019). Global energy transformation: A roadmap to 2050. International Renewable Energy Agency.
7. UNEP. (2019). Global trends in renewable energy investment 2019. UN Environment Programme.
8. OECD. (2017). OECD digital economy outlook 2017. OECD Publishing.
9. Schwab, K. (2017). The fourth industrial revolution. Crown Business.
10. UNESCO. (2015). UNESCO science report: Towards 2030. UNESCO Publishing.
11. Soete, L., Verspagen, B., & Ter Weel, B. (2010). Systems of innovation. In Handbook of the Economics of Innovation (Vol. 2, pp. 1159-1180). North-Holland.
12. Guellec, D., & Van Pottelsberghe de la Potterie, B. (2004). From R&D to productivity growth: Do the institutional settings and the source of funds of R&D matter? Oxford Bulletin of Economics and Statistics, 66(3), 353-378.
13. Acemoglu, D., Akcigit, U., Alp, H., Bloom, N., & Kerr, W. (2018). Innovation, reallocation, and growth. American Economic Review, 108(11), 3450-91.

14. Etzkowitz, H. (2008). The triple helix: University-industry-government innovation in action. Routledge.
15. Dutta, S., Lanvin, B., & Wunsch-Vincent, S. (2019). Global innovation index 2019: Creating healthy lives—the future of medical innovation. WIPO.
16. Feld, B. (2020). Startup communities: Building an entrepreneurial ecosystem in your city. John Wiley & Sons.
17. World Economic Forum. (2020). The future of jobs report 2020. World Economic Forum.
18. OECD. (2018). Education at a glance 2018: OECD indicators. OECD Publishing.
19. Spelt, E. J., Biemans, H. J., Tobi, H., Luning, P. A., & Mulder, M. (2009). Teaching and learning in interdisciplinary higher education: A systematic review. Educational Psychology Review, 21(4), 365-378.
20. Edler, J., & Fagerberg, J. (2017). Innovation policy: What, why, and how. Oxford Review of Economic Policy, 33(1), 2-23.

The Impact of Scientific Illiteracy on America's Workforce and Competitiveness

In an increasingly knowledge-based and technology-driven global economy, scientific literacy has become a critical determinant of a nation's workforce quality and competitive advantage. The United States, once a leader in science and innovation, now faces a growing challenge of scientific illiteracy among its population, which threatens to undermine its economic competitiveness and future prosperity.

Scientific literacy encompasses not only a basic understanding of scientific concepts and processes but also the ability to apply this knowledge to real-world problems, think critically about scientific information, and make informed decisions based on evidence [1]. These skills are becoming increasingly important across a wide range of industries and occupations, from healthcare and energy to manufacturing and agriculture [2].

However, recent studies have revealed a troubling lack of scientific literacy among the American public. A 2018 survey by the Pew Research Center found that only 52% of American adults could correctly answer a series of basic science questions, such as whether antibiotics can kill viruses or whether Earth's core is hot [3]. Similarly, the National Assessment of Educational Progress (NAEP) has consistently shown that a majority of U.S. students fail to achieve proficiency in science at the elementary, middle, and high school levels [4].

This lack of scientific literacy has significant implications for the quality and competitiveness of America's workforce. In today's economy, many of the fastest-growing and highest-paying jobs require a strong foundation in science, technology, engineering, and mathematics (STEM) [5]. These include occupations such as software developers, data analysts, biomedical engineers, and renewable energy technicians, which are critical for driving innovation and economic growth [6].

However, the U.S. is facing a growing shortage of skilled workers in these fields, due in part to the inadequate preparation of students in STEM subjects. A 2018 report by the National Science Board found that the U.S. is falling behind other nations in the production of STEM degrees, particularly at the graduate level [7]. This shortage of STEM talent is forcing many American companies to rely on foreign workers to fill critical positions, which can limit their ability to innovate and compete globally [8].

Moreover, scientific illiteracy can hinder the ability of American workers to adapt to the rapidly changing demands of the 21st-century workplace. As new technologies and industries emerge, workers will need to continually update their skills and knowledge to remain competitive [9]. This requires not only a strong foundation in STEM subjects but also the ability to learn and apply new scientific concepts throughout one's career [10].

The impact of scientific illiteracy on America's workforce and competitiveness is not limited to STEM fields alone. Many other industries, such as healthcare, agriculture, and manufacturing, also rely heavily on scientific knowledge and innovation [11]. For example, healthcare workers need to understand the basic principles of biology, chemistry, and physiology to provide effective patient care and keep up with the latest medical advancements [12]. Similarly, farmers need to have a strong grasp of plant science, soil science, and environmental science to optimize crop yields and manage natural resources sustainably [13].

Furthermore, scientific illiteracy can undermine the ability of American businesses to make informed decisions and investments in research and development (R&D). In today's global economy,

R&D is a critical driver of innovation and competitive advantage, allowing companies to develop new products, processes, and services that meet the evolving needs of customers and markets [14]. However, if business leaders and managers lack a basic understanding of scientific principles and methods, they may struggle to evaluate the potential risks and rewards of R&D investments, leading to suboptimal decision-making and resource allocation [15].

The consequences of scientific illiteracy for America's economic competitiveness are far-reaching and potentially severe. In a world where knowledge and innovation are increasingly the keys to success, nations that fail to cultivate a scientifically literate workforce risk falling behind in the global race for talent, investment, and market share [16]. This, in turn, can lead to slower economic growth, reduced living standards, and diminished influence on the world stage [17].

To address this challenge, the U.S. must make a concerted effort to improve scientific literacy and STEM education at all levels, from early childhood through adult learning [18]. This will require a comprehensive and coordinated approach that involves policymakers, educators, businesses, and communities working together to create a culture that values and supports scientific knowledge and innovation [19].

One key strategy is to invest in high-quality STEM education programs that engage students in hands-on, inquiry-based learning experiences and provide them with the skills and knowledge needed to succeed in the 21st-century workforce [20]. This may involve updating curricula and teaching methods to reflect the latest scientific advancements, providing teachers with the training and resources they need to deliver effective instruction, and creating partnerships between schools, universities, and industry to expose students to real-world applications of science and technology [21].

Another important approach is to promote lifelong learning and skill development among adult workers, particularly those in industries undergoing rapid technological change [22]. This may involve offering targeted training programs, apprenticeships, and certification courses that help workers acquire new scientific and technical skills, as well as providing incentives for employers to invest in the continuous learning and development of their workforce [23].

Ultimately, addressing the impact of scientific illiteracy on America's workforce and competitiveness will require a long-term commitment to building a scientifically literate society that values and supports innovation and discovery. By investing in STEM education, promoting lifelong learning, and fostering a culture of scientific inquiry and entrepreneurship, the U.S. can position itself to lead the world in the knowledge economy of the 21st century and ensure a brighter, more prosperous future for all its citizens.

References

1. National Academies of Sciences, Engineering, and Medicine. (2016). Science literacy: Concepts, contexts, and consequences. National Academies Press.
2. Bughin, J., Hazan, E., Lund, S., Dahlström, P., Wiesinger, A., & Subramaniam, A. (2018). Skill shift: Automation and the future of the workforce. McKinsey Global Institute.
3. Pew Research Center. (2018). What Americans know about science.
4. National Center for Education Statistics. (2019). The nation's report card: Science 2019.
5. Bureau of Labor Statistics. (2020). Employment projections: 2019-2029. U.S. Department of Labor.
6. Deming, D. J., & Noray, K. (2020). Earnings dynamics, changing job skills, and STEM careers. The Quarterly Journal of Economics, 135(4), 1965-2005.
7. National Science Board. (2018). Science and engineering indicators 2018. National Science Foundation.
8. Hanson, G. H., & Slaughter, M. J. (2017). High-skilled immigration and the rise of STEM occupations in US employment. In Education, skills, and technical change: Implications for future US GDP growth (pp. 465-494). University of Chicago Press.
9. World Economic Forum. (2020). The future of jobs report 2020. World Economic Forum.
10. National Research Council. (2012). Education for life and work: Developing transferable knowledge and skills in the 21st century. National Academies Press.
11. Beede, D., Julian, T., Langdon, D., McKittrick, G., Khan, B., & Doms, M. (2011). Women in STEM: A gender gap to innovation. U.S. Department of Commerce, Economics and Statistics Administration.
12. Frenk, J., Chen, L., Bhutta, Z. A., Cohen, J., Crisp, N., Evans, T., ... & Zurayk, H. (2010). Health professionals for a new century: Transforming education to strengthen health systems in an interdependent world. The Lancet, 376(9756), 1923-1958.
13. National Research Council. (2009). Transforming agricultural education for a changing world. National Academies Press.
14. Soete, L., Verspagen, B., & Ter Weel, B. (2010). Systems of innovation. In Handbook of the Economics of Innovation (Vol. 2, pp. 1159-1180). North-Holland.
15. Leiponen, A. (2005). Skills and innovation. International Journal of Industrial Organization, 23(5-6), 303-323.
16. OECD. (2012). Innovation for development: A discussion of the issues and an overview of work of the OECD directorate for science, technology and industry. OECD Publishing.
17. Acemoglu, D., & Restrepo, P. (2018). Artificial intelligence, automation and work. In The economics of artificial intelligence: An agenda (pp. 197-236). University of Chicago Press.
18. NASEM. (2020). Building capacity for teaching engineering in K-12 education. National Academies Press.

19. PCAST. (2010). Prepare and inspire: K-12 education in science, technology, engineering, and math (STEM) for America's future. Executive Office of the President.
20. Henderson, C., Beach, A., & Finkelstein, N. (2011). Facilitating change in undergraduate STEM instructional practices: An analytic review of the literature. Journal of Research in Science Teaching, 48(8), 952-984.
21. Duff, J. M., & Mellis, A. M. (2020). Best practices for industry engagement in STEM education. Journal of STEM Education, 21(1), 25-32.
22. Gauri, V., & Vawda, A. (2004). Vouchers for basic education in developing economies: A principal-agent perspective. World Bank Policy Research Working Paper 3005.
23. OECD. (2019). Getting skills right: Future-ready adult learning systems. OECD Publishing.

Case Studies: Declining Interest in STEM Fields and Brain Drain

The impact of scientific illiteracy on America's workforce and competitiveness can be vividly illustrated through case studies that highlight the real-world consequences of the growing disconnect between science and society. Two particularly pressing issues are the declining interest in STEM fields among American students and the phenomenon of "brain drain," whereby talented scientists and engineers leave the U.S. for better opportunities abroad.

The declining interest in STEM fields among American students is a worrying trend that threatens to undermine the nation's future supply of skilled workers in critical industries. Despite the growing demand for STEM professionals in the job market, the number of students pursuing degrees in these fields has remained stagnant or even declined in recent years [1]. For example, between 2003 and 2015, the number of bachelor's degrees awarded in computer science in the U.S. fell by 10%, even as the total number of bachelor's degrees increased by 50% [2].

This trend is particularly pronounced among women and underrepresented minorities, who continue to be significantly underrepresented in STEM fields [3]. In 2018, women earned only 19% of bachelor's degrees in computer science and 21% in engineering, while Black and Hispanic students earned just 10% and 13% of STEM degrees, respectively [4]. This lack of diversity not only limits the talent pool available to American companies but also hinders the development of innovative solutions that benefit from a wide range of perspectives and experiences [5].

Several factors contribute to the declining interest in STEM fields among American students. One major factor is the inadequate preparation and support provided by the K-12 education system, which often fails to cultivate a strong foundation in math and science or to provide students with engaging, hands-on learning experiences that spark their curiosity and interest [6]. Another factor is the persistence of negative stereotypes and cultural attitudes that portray STEM fields as difficult, boring, or unsuitable for certain groups, such as women and minorities [7].

The consequences of this declining interest in STEM are far-reaching and potentially severe. Without a robust pipeline of skilled workers in these fields, American companies may struggle to fill critical positions, leading to slower innovation, reduced competitiveness, and lost opportunities for economic growth [8]. Moreover, the lack of diversity in STEM can perpetuate inequalities and limit the nation's ability to tap into the full potential of its human capital [9].

The phenomenon of "brain drain" poses another significant challenge to America's technological advancement and economic competitiveness. Brain drain refers to the migration of highly skilled workers, particularly in STEM fields, from one country to another in search of better opportunities, higher salaries, or more supportive research environments [10]. For the U.S., which has long relied on its ability to attract and retain top talent from around the world, brain drain represents a growing threat to its position as a global leader in science and innovation.

In recent years, there have been increasing reports of American scientists and engineers leaving the country to pursue careers abroad, particularly in emerging economies such as China and India [11]. This trend is driven by a combination of factors, including the lack of stable funding for scientific research in the U.S., the growing competition for talent from other countries, and the perception that the U.S. is becoming less welcoming to international students and workers [12].

The impact of brain drain on America's workforce and competitiveness can be significant and long-lasting. When talented scientists and engineers leave the country, they take with them not only their skills and knowledge but also their networks, ideas, and potential contributions to American society [13]. This loss of human capital can slow down the pace of innovation, reduce the diversity of perspectives in research and development, and weaken the nation's ability to tackle complex challenges in fields such as healthcare, energy, and national security [14].

Moreover, brain drain can create a self-reinforcing cycle, whereby the departure of top talent makes it harder for American institutions to attract and retain the next generation of scientists and engineers [15]. This can lead to a gradual erosion of the nation's scientific and technological capabilities, as well as its global standing and influence [16].

To address the challenges of declining interest in STEM and brain drain, the U.S. will need to take a comprehensive and proactive approach that involves multiple stakeholders, including policymakers, educators, businesses, and communities [17]. This may include initiatives such as:

- Investing in high-quality STEM education programs that provide students with engaging, relevant, and hands-on learning experiences from an early age [18].
- Promoting diversity and inclusion in STEM fields by addressing systemic barriers, challenging stereotypes, and providing mentorship and support for underrepresented groups [19].
- Increasing funding for scientific research and development, particularly in areas of strategic importance to the nation's long-term competitiveness [20].
- Creating more attractive and supportive career pathways for scientists and engineers, including stable job opportunities, competitive salaries, and opportunities for professional growth and leadership [21].
- Fostering a culture of innovation and entrepreneurship that values and rewards scientific talent, and that encourages collaboration between academia, industry, and government [22].

Ultimately, addressing the declining interest in STEM and brain drain will require a sustained commitment to building a strong and inclusive scientific community that can attract, develop, and retain the best and brightest minds from across the country and around the world. By investing in the human capital that drives innovation and discovery, the U.S. can position itself to lead the way in the technological and economic transformations of the 21st century.

References

1. National Science Board. (2020). Science and engineering indicators 2020. National Science Foundation.
2. National Academies of Sciences, Engineering, and Medicine. (2018). Assessing and responding to the growth of computer science undergraduate enrollments. National Academies Press.
3. National Science Foundation. (2019). Women, minorities, and persons with disabilities in science and engineering. National Center for Science and Engineering Statistics.
4. National Center for Education Statistics. (2020). Digest of education statistics. U.S. Department of Education.
5. Page, S. E. (2007). The difference: How the power of diversity creates better groups, firms, schools, and societies. Princeton University Press.
6. National Research Council. (2011). Successful K-12 STEM education: Identifying effective approaches in science, technology, engineering, and mathematics. National Academies Press.
7. Cheryan, S., Ziegler, S. A., Montoya, A. K., & Jiang, L. (2017). Why are some STEM fields more gender balanced than others? Psychological Bulletin, 143(1), 1-35.
8. Rothwell, J. (2013). The hidden STEM economy. Brookings Institution.
9. Ong, M., Smith, J. M., & Ko, L. T. (2018). Counterspaces for women of color in STEM higher education: Marginal and central spaces for persistence and success. Journal of Research in Science Teaching, 55(2), 206-245.
10. Docquier, F., & Rapoport, H. (2012). Globalization, brain drain, and development. Journal of Economic Literature, 50(3), 681-730.
11. Zweig, D., & Wang, H. (2013). Can China bring back the best? The Communist Party organizes China's search for talent. The China Quarterly, 215, 590-615.
12. Hanson, G. H., & Slaughter, M. J. (2016). High-skilled immigration and the rise of STEM occupations in US employment. NBER Working Paper No. 22623.
13. Kahn, S., & MacGarvie, M. (2016). How important is U.S. location for research in science? The Review of Economics and Statistics, 98(2), 397-414.
14. National Science Board. (2020). Vision 2030. National Science Foundation.
15. Stephan, P. E. (2012). How economics shapes science. Harvard University Press.
16. Freeman, R. B., & Huang, W. (2015). China's "Great Leap Forward" in science and engineering. In A. Geuna (Ed.), Global mobility of research scientists: The economics of who goes where and why (pp. 155-175). Academic Press.
17. National Academies of Sciences, Engineering, and Medicine. (2007). Rising above the gathering storm: Energizing and employing America for a brighter economic future. National Academies Press.
18. Kelley, T. R., & Knowles, J. G. (2016). A conceptual framework for integrated STEM education. International Journal of STEM Education, 3(1), 1-11.
19. Estrada, M., Burnett, M., Campbell, A. G., Campbell, P. B., Denetclaw, W. F., Gutiérrez, C. G., ... & Zavala, M. (2016). Improving underrepresented minority student persistence in STEM. CBE—Life Sciences Education, 15(3), es5.

20. American Academy of Arts and Sciences. (2014). Restoring the foundation: The vital role of research in preserving the American dream. American Academy of Arts and Sciences.
21. National Academies of Sciences, Engineering, and Medicine. (2018). Graduate STEM education for the 21st century. National Academies Press.
22. Etzkowitz, H. (2008). The triple helix: University-industry-government innovation in action. Routledge.

Chapter 7: Political Decision-Making and Democracy

The Role of Science in Informing Policy Decisions

In a well-functioning democracy, political decision-making should be guided by reliable information, sound reasoning, and a thorough understanding of the complex issues at hand. Science, with its systematic approach to gathering and analyzing evidence, plays a crucial role in informing policy decisions across a wide range of domains, from public health and environmental protection to national security and economic development.

At its core, science is a process of inquiry that seeks to generate reliable knowledge about the natural and social world [1]. By formulating hypotheses, collecting data, and subjecting ideas to rigorous testing and peer review, scientists aim to develop a robust understanding of the phenomena they study, free from bias, ideology, or personal interest [2]. This commitment to objectivity and empirical evidence makes science an invaluable resource for policymakers, who are often called upon to make difficult decisions in the face of complex and uncertain challenges.

One of the most important ways in which science informs policy is by providing a factual basis for understanding the nature and scope of public problems. Through research and analysis, scientists can help to identify the underlying causes of societal challenges, assess the magnitude and distribution of their impacts, and forecast their likely future trajectories [3]. This information is essential for policymakers, as it allows them to prioritize issues, allocate resources, and develop targeted interventions that are grounded in evidence rather than assumptions or political expediency.

For example, in the realm of public health, epidemiological studies have been instrumental in identifying the risk factors and transmission pathways of infectious diseases, informing the development of effective prevention and control strategies [4]. Similarly, in the field of environmental policy, scientific research on the causes and consequences of climate change has helped to build a case for urgent action to reduce greenhouse gas emissions and adapt to the impacts of a warming planet [5].

Science also plays a critical role in evaluating the effectiveness and efficiency of policy interventions. By designing and conducting rigorous studies, scientists can assess the outcomes of policies and programs, identify their strengths and weaknesses, and suggest ways to improve their performance [6]. This feedback loop between policy and research is essential for ensuring that public resources are used wisely and that policies are continually refined and adapted to changing circumstances.

In the field of education, for instance, scientific research has been instrumental in identifying effective teaching practices and learning strategies, informing the design of curricula, assessments, and teacher training programs [7]. Similarly, in the realm of criminal justice, studies on the effectiveness of various policing strategies, sentencing policies, and rehabilitation programs have helped to guide reforms aimed at reducing crime, promoting public safety, and ensuring the fair administration of justice [8].

However, the relationship between science and policy is not always straightforward or harmonious. Scientists and policymakers often operate in different institutional contexts, with different incentives, constraints, and ways of knowing [9]. Scientists are trained to be cautious, tentative, and transparent about the limitations of their research, while policymakers are often under pressure to make quick decisions based on incomplete information and competing political priorities [10].

Moreover, science is not always able to provide definitive answers to policy questions, especially in areas where the evidence is complex, uncertain, or contested. In such cases, policymakers may need to make judgments based on values, interests, and other considerations that go beyond the scope of scientific analysis [11]. This is particularly true in areas where policy decisions involve trade-offs between different social goals or where the consequences of action (or inaction) are not fully known.

The challenges of integrating science into policy are compounded by the fact that science is not a monolithic or infallible enterprise. Like any human endeavor, science is subject to social, cultural, and political influences that can shape the questions researchers ask, the methods they use, and the conclusions they draw [12]. Scientists may disagree about the interpretation of data, the validity of methods, or the implications of findings, leading to ongoing debates and uncertainties that can be difficult for policymakers to navigate [13].

To ensure that science effectively informs policy decisions, it is essential to foster a culture of open and honest communication between researchers and policymakers. This requires building trust, mutual understanding, and a shared commitment to the public good [14]. Scientists must be willing to engage with policymakers and the public, to communicate their findings in clear and accessible ways, and to be transparent about the limitations and uncertainties of their research [15]. Policymakers, in turn, must be willing to seek out and consider scientific evidence, to ask critical questions, and to make decisions based on the best available information, even when it may be politically inconvenient [16].

Fostering effective science-policy interactions also requires institutional support and resources. Governments and other organizations can play a key role in promoting the use of science in policy by providing funding for policy-relevant research, creating mechanisms for science advice and consultation, and building capacity for science communication and public engagement [17]. Universities and research institutions can also contribute by training scientists to work at the interface of science and policy, and by creating incentives and rewards for researchers who engage in policy-relevant work [18].

Ultimately, the role of science in informing policy decisions is essential for promoting evidence-based decision-making and for ensuring that public policies are effective, efficient, and responsive to the needs of society. By working together to build a strong and mutually reinforcing relationship between science and policy, we can create a more informed, engaged, and democratic society that is better equipped to tackle the complex challenges of the 21st century.

References

1. National Academies of Sciences, Engineering, and Medicine. (2017). Communicating science effectively: A research agenda. National Academies Press.
2. Popper, K. (1959). The logic of scientific discovery. Routledge.
3. Nutley, S., Walter, I., & Davies, H. (2007). Using evidence: How research can inform public services. Policy Press.
4. Brownson, R. C., Fielding, J. E., & Maylahn, C. M. (2009). Evidence-based public health: A fundamental concept for public health practice. Annual Review of Public Health, 30, 175-201.
5. Intergovernmental Panel on Climate Change. (2014). Climate change 2014: Synthesis report. Contribution of Working Groups I, II and III to the fifth assessment report of the Intergovernmental Panel on Climate Change.
6. Weiss, C. H. (1979). The many meanings of research utilization. Public Administration Review, 39(5), 426-431.
7. Slavin, R. E. (2002). Evidence-based education policies: Transforming educational practice and research. Educational Researcher, 31(7), 15-21.
8. Sherman, L. W. (2013). The rise of evidence-based policing: Targeting, testing, and tracking. Crime and Justice, 42(1), 377-451.
9. Jasanoff, S. (1990). The fifth branch: Science advisers as policymakers. Harvard University Press.
10. Pielke Jr, R. A. (2007). The honest broker: Making sense of science in policy and politics. Cambridge University Press.
11. Douglas, H. (2009). Science, policy, and the value-free ideal. University of Pittsburgh Press.
12. Gieryn, T. F. (1999). Cultural boundaries of science: Credibility on the line. University of Chicago Press.
13. Sarewitz, D. (2004). How science makes environmental controversies worse. Environmental Science & Policy, 7(5), 385-403.
14. Cash, D. W., Clark, W. C., Alcock, F., Dickson, N. M., Eckley, N., Guston, D. H., ... & Mitchell, R. B. (2003). Knowledge systems for sustainable development. Proceedings of the National Academy of Sciences, 100(14), 8086-8091.
15. Baron, N. (2010). Escape from the ivory tower: A guide to making your science matter. Island Press.
16. Cairney, P. (2016). The politics of evidence-based policy making. Springer.
17. Gluckman, P. (2014). Policy: The art of science advice to government. Nature, 507(7491), 163-165.
18. Petes, L. E., & Meyer, M. D. (2018). An ecologist's guide to careers in science policy advising. Frontiers in Ecology and the Environment, 16(1), 53-54.

The Impact of Scientific Illiteracy on Voter Choices and Political Discourse

In a democratic society, the choices made by voters and the quality of political discourse play a crucial role in shaping public policies and the direction of the nation. However, when a significant portion of the electorate lacks a basic understanding of scientific principles and methods, it can have profound consequences for the functioning of democracy and the ability of society to make informed decisions about complex issues.

One of the most direct ways in which scientific illiteracy impacts voter choices is by making individuals more susceptible to misinformation and manipulation. In an era of social media and 24-hour news cycles, citizens are constantly bombarded with information from a wide range of sources, not all of which are reliable or accurate [1]. When voters lack the scientific knowledge and critical thinking skills necessary to evaluate the credibility of this information, they may be more likely to fall for false or misleading claims, especially those that align with their preexisting beliefs or biases [2].

This vulnerability to misinformation can have serious consequences for the political process. Voters who are misinformed about key issues may make choices that are contrary to their own interests or values, or that have negative impacts on society as a whole [3]. For example, a lack of understanding about the science of climate change may lead voters to support candidates or policies that prioritize short-term economic gains over long-term environmental sustainability, even if they personally value protecting the planet for future generations [4].

Scientific illiteracy can also contribute to the polarization and tribalization of political discourse. When voters lack a shared understanding of basic scientific facts and methods, they may be more likely to retreat into ideological echo chambers, where they are exposed only to information that confirms their existing beliefs [5]. This can lead to a breakdown in communication and compromise across political divides, as well as an erosion of trust in democratic institutions and processes [6].

Moreover, when political leaders and media outlets exploit scientific illiteracy for political gain, it can further undermine the quality of public discourse. Politicians who dismiss or distort scientific evidence in order to appeal to certain constituencies or advance their own agendas contribute to a climate of confusion and mistrust, making it harder for voters to make informed decisions [7]. Similarly, media outlets that prioritize sensationalism and conflict over accuracy and nuance can reinforce misconceptions and fuel polarization, rather than promoting a shared understanding of complex issues [8].

The impact of scientific illiteracy on voter choices and political discourse is particularly concerning in an era of global challenges that require evidence-based solutions. From climate change and pandemics to artificial intelligence and cybersecurity, many of the most pressing issues facing society today have significant scientific dimensions that demand a well-informed and engaged citizenry [9]. When large segments of the population lack the knowledge and skills necessary to participate meaningfully in these debates, it can lead to suboptimal policies and missed opportunities for progress.

To address these challenges, it is essential to prioritize scientific literacy as a core component of civic education and engagement. This means not only teaching students the basic facts and concepts of science but also equipping them with the critical thinking and media literacy skills necessary to navigate the complex information landscape of the 21st century [10]. It also means creating more opportunities for dialogue and collaboration between scientists, policymakers, and the public, to build trust and understanding across different perspectives and backgrounds [11].

One promising approach is to focus on science communication and public engagement initiatives that meet people where they are, using language, formats, and channels that are accessible and relevant to diverse audiences [12]. This may involve leveraging social media and other digital platforms to reach younger generations, partnering with community organizations and influencers to build trust and credibility, and using storytelling and other narrative techniques to make complex scientific ideas more relatable and compelling [13].

Another key strategy is to empower citizens to participate more actively in the scientific process itself, through citizen science projects, public consultations, and other forms of participatory research [14]. By involving the public in the co-creation of scientific knowledge and the development of solutions to societal challenges, we can foster a greater sense of ownership and investment in the outcomes of science, as well as a deeper appreciation for the methods and values of scientific inquiry [15].

Ultimately, addressing the impact of scientific illiteracy on voter choices and political discourse will require a sustained and multi-faceted effort that involves educators, policymakers, scientists, and citizens working together to build a more scientifically literate and engaged society. By prioritizing the development of scientific knowledge and skills as a core component of civic life, we can create a more informed, resilient, and democratic society that is better equipped to tackle the complex challenges of the 21st century.

Only by cultivating a shared understanding of the value and methods of science, and by creating more opportunities for meaningful public engagement with scientific issues, can we hope to build a political culture that is more responsive to evidence, more resilient to misinformation, and more capable of making wise decisions for the common good. This will require a significant investment of resources and political will, but the stakes could not be higher. The future of our democracy, and our ability to solve the great challenges of our time, may well depend on our success in this endeavor.

References

1. Vosoughi, S., Roy, D., & Aral, S. (2018). The spread of true and false news online. Science, 359(6380), 1146-1151.
2. Scheufele, D. A., & Krause, N. M. (2019). Science audiences, misinformation, and fake news. Proceedings of the National Academy of Sciences, 116(16), 7662-7669.
3. Lewandowsky, S., Ecker, U. K., & Cook, J. (2017). Beyond misinformation: Understanding and coping with the "post-truth" era. Journal of Applied Research in Memory and Cognition, 6(4), 353-369.

4. Hornsey, M. J., Harris, E. A., Bain, P. G., & Fielding, K. S. (2016). Meta-analyses of the determinants and outcomes of belief in climate change. Nature Climate Change, 6(6), 622-626.
5. Bakshy, E., Messing, S., & Adamic, L. A. (2015). Exposure to ideologically diverse news and opinion on Facebook. Science, 348(6239), 1130-1132.
6. Iyengar, S., & Westwood, S. J. (2015). Fear and loathing across party lines: New evidence on group polarization. American Journal of Political Science, 59(3), 690-707.
7. Mooney, C. (2005). The Republican war on science. Basic Books.
8. Bennett, W. L., & Iyengar, S. (2008). A new era of minimal effects? The changing foundations of political communication. Journal of Communication, 58(4), 707-731.
9. National Academies of Sciences, Engineering, and Medicine. (2017). Communicating science effectively: A research agenda. National Academies Press.
10. Rudolph, J. L. (2020). Science education in the age of misinformation. Science Education, 104(6), 979-985.
11. Fischhoff, B. (2013). The sciences of science communication. Proceedings of the National Academy of Sciences, 110(Supplement 3), 14033-14039.
12. Nisbet, M. C., & Scheufele, D. A. (2009). What's next for science communication? Promising directions and lingering distractions. American Journal of Botany, 96(10), 1767-1778.
13. Dahlstrom, M. F. (2014). Using narratives and storytelling to communicate science with nonexpert audiences. Proceedings of the National Academy of Sciences, 111(Supplement 4), 13614-13620.
14. Bonney, R., Phillips, T. B., Ballard, H. L., & Enck, J. W. (2016). Can citizen science enhance public understanding of science? Public Understanding of Science, 25(1), 2-16.
15. Irwin, A. (2018). No PhDs needed: How citizen science is transforming research. Nature, 562(7728), 480-482.

Consequences for Evidence-Based Policymaking and Democracy

The erosion of scientific literacy and the spread of misinformation have far-reaching consequences that extend beyond individual voter choices and political discourse. When a significant portion of the public lacks the knowledge and skills necessary to engage critically with scientific evidence, it can undermine the very foundation of evidence-based policymaking and threaten the health of democratic institutions.

Evidence-based policymaking refers to the process of using the best available scientific evidence to inform and guide decisions about public policy [1]. This approach is grounded in the idea that policies should be based on a thorough understanding of the facts, rather than on ideology, intuition, or political expediency. By drawing on the expertise of scientists and other experts, policymakers can craft solutions that are more likely to be effective, efficient, and responsive to the needs of society [2].

However, when scientific illiteracy is widespread, it can create significant barriers to evidence-based policymaking. Policymakers who lack a basic understanding of scientific concepts and methods may struggle to interpret and apply scientific evidence in their decision-making processes. They may be more likely to rely on anecdotal evidence, personal beliefs, or the influence of special interest groups, rather than on rigorous data and analysis [3].

Moreover, when the public is not equipped to evaluate the credibility of scientific claims or to distinguish between reliable and unreliable sources of information, it can create a permissive environment for the spread of misinformation and pseudoscience. This can lead to a situation where policymakers feel pressure to respond to public demands that are based on false or misleading information, rather than on evidence [4].

The consequences of this breakdown in evidence-based policymaking can be severe. Policies that are not grounded in science may be ineffective at best and harmful at worst. They may waste public resources, exacerbate social inequalities, or have unintended consequences that could have been avoided with a more evidence-based approach [5].

For example, in the realm of public health, policies that are based on misinformation or pseudoscience can have devastating consequences. The anti-vaccination movement, which has spread false claims about the safety and efficacy of vaccines, has led to declining vaccination rates and the resurgence of preventable diseases in many communities [6]. Similarly, the promotion of unproven or dangerous treatments for COVID-19, such as hydroxychloroquine or ivermectin, has led some people to reject evidence-based public health measures and put themselves and others at risk [7].

In the realm of environmental policy, a lack of scientific literacy can lead to a failure to act on urgent threats such as climate change. When large segments of the public and political leaders dismiss or downplay the overwhelming scientific evidence of human-caused global warming, it becomes much more difficult to enact policies that would reduce greenhouse gas emissions and mitigate the worst impacts of climate change [8]. This delay in action can have catastrophic consequences for future generations and for the planet as a whole.

Beyond specific policy areas, the erosion of evidence-based policymaking can also have broader implications for the functioning of democratic institutions. When the public loses trust in the ability of scientists and other experts to provide reliable and objective information, it can lead to a more general erosion of trust in government and other institutions [9].

This loss of trust can create a vicious cycle, where the public becomes increasingly skeptical of expert advice and more vulnerable to misinformation and conspiracy theories. This, in turn, makes it harder for policymakers to enact evidence-based policies, as they face greater public resistance and political pressure to cater to popular opinion, even when that opinion is not grounded in facts [10].

Over time, this dynamic can lead to a situation where the very notion of objective truth becomes contested and where political debates devolve into a battle of competing narratives, rather than a good-faith effort to find common ground based on shared facts [11]. This kind of "post-truth" politics is deeply corrosive to democratic norms and institutions, as it undermines the ability of citizens to hold their leaders accountable and to participate meaningfully in the political process [12].

To address these challenges, it is essential to prioritize scientific literacy and critical thinking skills as core components of citizenship and civic engagement. This means investing in science education at all levels, from K-12 schools to adult learning programs, and creating more opportunities for the public to engage directly with scientists and scientific institutions [13].

It also means working to rebuild trust in science and expertise, by increasing transparency and accountability in the scientific process, and by creating more opportunities for scientists to communicate directly with the public and policymakers [14]. This may involve new models of science communication and engagement, such as citizen science initiatives, deliberative forums, and partnerships with community organizations and media outlets [15].

Importantly, efforts to promote evidence-based policymaking must also grapple with the broader social and political factors that can shape how science is used (or misused) in the policy process. This includes addressing issues of power, inequality, and special interests, and working to ensure that the production and use of scientific evidence is guided by democratic values such as transparency, inclusion, and accountability [16].

Ultimately, the consequences of scientific illiteracy for evidence-based policymaking and democracy are too significant to ignore. By prioritizing scientific literacy and critical thinking skills as essential components of civic life, and by working to build a culture of evidence-based decision-making that is responsive to the needs and values of diverse communities, we can create a stronger, more resilient democracy that is better equipped to tackle the complex challenges of the 21st century.

The alternative – a society where facts are treated as mere opinions, where expertise is dismissed as elitism, and where public policy is driven by fear, prejudice, and misinformation – is a path that leads only to dysfunction, discord, and decline. It is up to all of us – as citizens, educators, scientists, and policymakers – to chart a different course, one that upholds the vital role of science in a thriving democracy.

References

1. Cairney, P. (2016). The politics of evidence-based policy making. Springer.
2. Head, B. W. (2008). Three lenses of evidence-based policy. Australian Journal of Public Administration, 67(1), 1-11.
3. Parkhurst, J. (2017). The politics of evidence: From evidence-based policy to the good governance of evidence. Routledge.
4. Dietz, T. (2013). Bringing values and deliberation to science communication. Proceedings of the National Academy of Sciences, 110(Supplement 3), 14081-14087.
5. Cairney, P., Oliver, K., & Wellstead, A. (2016). To bridge the divide between evidence and policy: Reduce ambiguity as much as uncertainty. Public Administration Review, 76(3), 399-402.
6. Larson, H. J. (2018). The biggest pandemic risk? Viral misinformation. Nature, 562(7727), 309-309.
7. Aquino, Y. S. J., Cabrera, N., & Altubar, K. (2021). The "infodemic" of COVID-19: The fight against the manipulation of scientific evidence. JAMA, 325(14), 1371-1372.

8. Oreskes, N., & Conway, E. M. (2011). Merchants of doubt: How a handful of scientists obscured the truth on issues from tobacco smoke to global warming. Bloomsbury Publishing USA.
9. Gauchat, G. (2012). Politicization of science in the public sphere: A study of public trust in the United States, 1974 to 2010. American Sociological Review, 77(2), 167-187.
10. Nichols, T. (2017). The death of expertise: The campaign against established knowledge and why it matters. Oxford University Press.
11. Kavanagh, J., & Rich, M. D. (2018). Truth decay: An initial exploration of the diminishing role of facts and analysis in American public life. Rand Corporation.
12. Suiter, J. (2016). Post-truth politics. Political Insight, 7(3), 25-27.
13. Feinstein, N. W., & Waddington, D. I. (2020). Individual truth judgments or purposeful, collective sensemaking? Rethinking science education's response to the post-truth era. Educational Psychologist, 55(3), 155-166.
14. Jamieson, K. H., McNutt, M., Kiermer, V., & Sever, R. (2019). Signaling the trustworthiness of science. Proceedings of the National Academy of Sciences, 116(39), 19231-19236.
15. Safford, T. G., Whitmore, E. H., & Hamilton, L. C. (2020). The Carsey Perspective: Ideology Affects Trust in Science Agencies During a Pandemic (Issue Brief No. 22). Carsey School of Public Policy, University of New Hampshire.
16. Jasanoff, S. (2012). Science and public reason. Routledge.

Part III:
Addressing the Problem

Chapter 8: Improving Science Education

Strategies for Enhancing K-12 Science Curricula and Teaching Methods

Improving science education is a critical step in addressing the scientific literacy crisis and preparing students for the challenges and opportunities of the 21st century. At the heart of this effort is the need to transform K-12 science curricula and teaching methods in ways that engage students, foster critical thinking skills, and promote a deep understanding of scientific concepts and practices.

One key strategy for enhancing K-12 science curricula is to emphasize the integration of scientific practices with content knowledge. In the past, science education often focused primarily on memorizing facts and formulas, with little attention paid to the processes of scientific inquiry and problem-solving [1]. However, research has shown that students learn science more effectively when they engage in authentic scientific practices, such as asking questions, designing experiments, analyzing data, and constructing explanations [2].

To support this approach, science curricula should be designed around core disciplinary ideas and crosscutting concepts that provide a framework for understanding the natural world [3]. These ideas and concepts should be revisited and built upon throughout the K-12 curriculum, allowing students to develop a deep and interconnected understanding of science over time [4]. Additionally, curricula should incorporate real-world examples and problems that demonstrate the relevance and applicability of scientific knowledge to students' lives and communities [5].

Another important strategy is to adopt evidence-based teaching methods that have been shown to be effective in promoting student learning and engagement. One such method is inquiry-based learning, which involves guiding students through the process of scientific investigation, from posing questions to drawing conclusions based on evidence [6]. This approach can help students develop critical thinking skills, learn to work collaboratively, and take ownership of their own learning [7].

Other effective teaching methods include project-based learning, where students work on extended, interdisciplinary projects that integrate science with other subjects such as mathematics, engineering, and technology [8]; and problem-based learning, where students work in teams to solve complex, real-world problems using scientific knowledge and skills [9]. These approaches can help students see the connections between science and other fields, and develop the skills and mindsets needed for success in the modern workforce [10].

Technology can also play a transformative role in enhancing science education. Digital tools and resources, such as online simulations, virtual laboratories, and data visualization software, can provide students with access to authentic scientific experiences and data sets that would otherwise be difficult or impossible to replicate in the classroom [11]. These tools can also support personalized and adaptive learning, allowing students to progress at their own pace and receive targeted feedback and support [12].

However, integrating technology into science education is not a panacea, and must be done thoughtfully and strategically. Teachers need ongoing professional development and support to effectively use technology in their instruction, and to ensure that it enhances rather than replaces hands-on, inquiry-based learning [13]. Additionally, issues of access and equity must be addressed to ensure that all students, regardless of their socioeconomic status or geographic location, have the opportunity to benefit from technology-enhanced science education [14].

Improving science education also requires a focus on diversity, equity, and inclusion. Historically, science education has often excluded or marginalized students from underrepresented groups, such as women, minorities, and students with disabilities [15]. This not only limits the talent pool for future scientific and technological innovation but also reinforces stereotypes and biases that can discourage students from pursuing careers in science [16].

To address these issues, science curricula and teaching methods must be culturally responsive and relevant to the diverse experiences and backgrounds of all students [17]. This may involve incorporating examples and perspectives from a range of cultures and communities, highlighting the contributions of scientists from underrepresented groups, and creating inclusive classroom environments that value and respect all students' ideas and experiences [18].

Additionally, efforts must be made to provide equitable access to high-quality science education for all students, regardless of their race, ethnicity, gender, or socioeconomic status. This may involve targeted interventions and support for students who have been traditionally underserved by science education, such as providing mentoring and research opportunities, offering advanced coursework and enrichment programs, and partnering with community organizations and institutions to provide authentic science experiences outside of the classroom [19].

Ultimately, enhancing K-12 science curricula and teaching methods will require a sustained and collaborative effort from educators, policymakers, scientists, and communities. It will involve significant investments in teacher training and professional development, curricular reform and innovation, and the development and dissemination of high-quality instructional materials and resources [20].

However, the benefits of this effort are clear and compelling. By providing all students with a strong foundation in scientific knowledge and skills, and by fostering a love of learning and a passion for scientific inquiry, we can cultivate a new generation of informed, engaged, and empowered citizens who are prepared to tackle the complex challenges of the 21st century.

Science education is not just about preparing students for careers in science and technology, although that is certainly an important goal. It is also about developing the critical thinking skills, creativity, and problem-solving abilities that are essential for success in any field, and for active and informed citizenship in a rapidly changing world.

By embracing innovative and evidence-based approaches to science education, and by working to ensure that all students have access to high-quality learning opportunities, we can unlock the full potential of our nation's youth and build a brighter, more sustainable, and more equitable future for all.

References

1. National Research Council. (2012). A framework for K-12 science education: Practices, crosscutting concepts, and core ideas. National Academies Press.
2. Osborne, J. (2014). Teaching scientific practices: Meeting the challenge of change. Journal of Science Teacher Education, 25(2), 177-196.
3. National Academies of Sciences, Engineering, and Medicine. (2019). Science and engineering for grades 6-12: Investigation and design at the center. National Academies Press.
4. Corcoran, T., Mosher, F. A., & Rogat, A. (2009). Learning progressions in science: An evidence-based approach to reform. Consortium for Policy Research in Education.
5. Tal, T., & Kedmi, Y. (2006). Teaching socioscientific issues: Classroom culture and students' performances. Cultural Studies of Science Education, 1(4), 615-644.
6. Pedaste, M., Mäeots, M., Siiman, L. A., De Jong, T., Van Riesen, S. A., Kamp, E. T., ... & Tsourlidaki, E. (2015). Phases of inquiry-based learning: Definitions and the inquiry cycle. Educational Research Review, 14, 47-61.
7. Lazonder, A. W., & Harmsen, R. (2016). Meta-analysis of inquiry-based learning: Effects of guidance. Review of Educational Research, 86(3), 681-718.
8. Han, S., Capraro, R., & Capraro, M. M. (2015). How science, technology, engineering, and mathematics (STEM) project-based learning (PBL) affects high, middle, and low achievers differently: The impact of student factors on achievement. International Journal of Science and Mathematics Education, 13(5), 1089-1113.
9. Savery, J. R. (2015). Overview of problem-based learning: Definitions and distinctions. Essential readings in problem-based learning: Exploring and extending the legacy of Howard S. Barrows, 9, 5-15.
10. English, L. D. (2016). STEM education K-12: Perspectives on integration. International Journal of STEM Education, 3(1), 1-8.
11. Honey, M., & Hilton, M. (Eds.). (2011). Learning science through computer games and simulations. National Academies Press.
12. Peffer, M. E., Beckler, M. L., Schunn, C., Renken, M., & Revak, A. (2015). Science classroom inquiry (SCI) simulations: A novel method to scaffold science learning. PLOS One, 10(3), e0120638.
13. Sell, K. S., Herbert, B. E., Stuessy, C. L., & Schielack, J. (2006). Supporting student conceptual model development of complex earth systems through the use of multiple representations and inquiry. Journal of Geoscience Education, 54(3), 396-407.
14. Russo, M., Hecht, D., Burghardt, M. D., Hacker, M., & Saxman, L. (2011). Development of a multidisciplinary middle school mathematics infusion model. Middle Grades Research Journal, 6(2), 113-128.

15. Lee, O., & Buxton, C. A. (2010). Diversity and equity in science education: Research, policy, and practice. Teachers College Press.
16. Nasir, N. I. S., & Vakil, S. (2017). STEM-focused academies in urban schools: Tensions and possibilities. Journal of the Learning Sciences, 26(3), 376-406.
17. Brown, J. C. (2017). A metasynthesis of the complementarity of culturally responsive and inquiry-based science education in K-12 settings: Implications for advancing equitable science teaching and learning. Journal of Research in Science Teaching, 54(9), 1143-1173.
18. Bang, M., & Medin, D. (2010). Cultural processes in science education: Supporting the navigation of multiple epistemologies. Science Education, 94(6), 1008-1026.
19. Vakil, S., & Ayers, R. (2019). The racial politics of STEM education in the USA: Interrogations and explorations. Race Ethnicity and Education, 22(4), 449-458.
20. National Research Council. (2015). Identifying and supporting productive STEM programs in out-of-school settings. National Academies Press.

The Role of Universities and Informal Science Education

While K-12 schools play a critical role in providing the foundation for scientific literacy, universities and informal science education programs are essential for deepening and extending that knowledge throughout an individual's life. These institutions have the potential to engage learners of all ages in authentic scientific experiences, foster a lifelong love of learning, and prepare the next generation of scientists and innovators.

Universities, in particular, have a unique and vital role to play in advancing scientific literacy and preparing students for careers in science and technology. As the primary institutions responsible for training future scientists, engineers, and science educators, universities have the expertise and resources to provide students with cutting-edge knowledge and skills in their chosen fields [1]. Moreover, through their research and outreach activities, universities can contribute to the broader public understanding of science and its applications to society [2].

One way that universities can enhance science education is by providing opportunities for undergraduate students to engage in authentic research experiences. Participation in research has been shown to increase students' interest in science, deepen their understanding of scientific concepts and practices, and improve their critical thinking and problem-solving skills [3]. Many universities offer research programs, internships, and apprenticeships that allow students to work alongside faculty members and graduate students on real-world scientific projects [4].

In addition to research experiences, universities can also promote scientific literacy through their general education curricula. By requiring all students, regardless of their major, to take courses in the natural and social sciences, universities can help to ensure that all graduates have a basic understanding of scientific concepts and methods [5]. These courses can also provide opportunities for students to explore the connections between science and other disciplines, such as ethics, policy, and communication [6].

Beyond the classroom, universities can also engage the broader public in science through outreach and community engagement initiatives. Many universities have established science outreach programs that bring faculty and students into K-12 schools, community centers, and public events to share their knowledge and enthusiasm for science [7]. These programs can help to inspire and motivate young people to pursue careers in science, as well as to promote public understanding and support for scientific research [8].

Informal science education programs, such as museums, science centers, and after-school programs, also play a crucial role in promoting scientific literacy and engagement. These programs provide opportunities for learners of all ages to explore scientific concepts and phenomena in a hands-on, interactive way, outside of the structured classroom environment [9].

One of the key strengths of informal science education is its ability to reach diverse audiences and to promote equity and inclusion in science. Many informal science programs are designed to engage underrepresented groups, such as women, minorities, and low-income communities, who may not have access to high-quality science education in their schools [10]. By providing free or low-cost access to exhibits, workshops, and mentoring opportunities, these programs can help to level the playing field and inspire a love of science in all learners [11].

Informal science education programs can also provide a valuable complement to formal science education by offering a different perspective on scientific concepts and practices. While K-12 schools often focus on teaching basic scientific knowledge and skills, informal programs can delve deeper into specific topics and applications, such as environmental science, space exploration, or biotechnology [12]. Moreover, by emphasizing creativity, exploration, and discovery, informal programs can help to foster the kind of innovative thinking and problem-solving skills that are essential for success in the 21st century [13].

To maximize the impact of informal science education programs, it is important for them to collaborate and partner with universities, schools, and other community organizations. For example, many museums and science centers work closely with university researchers to develop exhibits and programs that showcase cutting-edge scientific research and its applications to society [14]. Similarly, after-school programs may partner with local schools to provide enrichment activities that reinforce and extend the science curriculum [15].

One promising model for collaboration between universities and informal science education programs is the "citizen science" approach, in which members of the public participate in real-world scientific research projects under the guidance of professional scientists [16]. Citizen science projects can range from monitoring local environmental conditions to analyzing astronomical data, and they provide opportunities for learners of all ages to contribute to scientific knowledge while developing their own skills and interests [17].

Ultimately, the role of universities and informal science education programs in promoting scientific literacy and engagement is both vital and complex. These institutions have the potential to inspire and empower learners of all ages and backgrounds to pursue their interests in science, to develop a deep understanding of scientific concepts and practices, and to apply that knowledge to real-world problems and challenges.

However, realizing this potential will require sustained investment, collaboration, and innovation from educators, researchers, policymakers, and community leaders. It will require a commitment to equity and inclusion, to ensure that all learners have access to high-quality science education experiences, both in and out of the classroom. And it will require a willingness to embrace new models and approaches, such as citizen science and interdisciplinary learning, that can help to break down the barriers between science and society.

By working together to strengthen and expand the role of universities and informal science education programs, we can create a more scientifically literate and engaged society, one that is better equipped to tackle the complex challenges of the 21st century. From climate change and public health to technological innovation and space exploration, the future of our world will depend on our ability to harness the power of science and to inspire the next generation of scientists and innovators. Universities and informal science education programs have a vital role to play in this effort, and their success will be essential for the prosperity and well-being of us all.

References

1. National Academies of Sciences, Engineering, and Medicine. (2016). Barriers and opportunities for 2-year and 4-year STEM degrees: Systemic change to support students' diverse pathways. National Academies Press.
2. Thiry, H., Laursen, S. L., & Hunter, A. B. (2011). What experiences help students become scientists? A comparative study of research and other sources of personal and professional gains for STEM undergraduates. The Journal of Higher Education, 82(4), 357-388.
3. Russell, S. H., Hancock, M. P., & McCullough, J. (2007). Benefits of undergraduate research experiences. Science, 316(5824), 548-549.
4. Lopatto, D. (2010). Undergraduate research as a high-impact student experience. Peer Review, 12(2), 27.
5. Olson, S., & Riordan, D. G. (2012). Engage to Excel: Producing One Million Additional College Graduates with Degrees in Science, Technology, Engineering, and Mathematics. Report to the President. Executive Office of the President.
6. Labov, J. B. (2004). From the National Academies: the challenges and opportunities for improving undergraduate science education through introductory courses. Cell Biology Education, 3(4), 212-214.
7. Moskal, B. M., Skokan, C., Kosbar, L., Dean, A., Westland, C., Barker, H., ... & Tafoya, J. (2007). K-12 outreach: Identifying the broader impacts of four outreach projects. Journal of Engineering Education, 96(3), 173-189.
8. Ufnar, J. A., Kuner, S., & Shepherd, V. L. (2012). Moving beyond GK-12. CBE—Life Sciences Education, 11(3), 239-247.

9. Dierking, L. D., Falk, J. H., Rennie, L., Anderson, D., & Ellenbogen, K. (2003). Policy statement of the "informal science education" ad hoc committee. Journal of Research in Science Teaching, 40(2), 108-111.
10. Dawson, E. (2014). "Not designed for us": How science museums and science centers socially exclude low-income, minority ethnic groups. Science Education, 98(6), 981-1008.
11. Bell, P., Lewenstein, B., Shouse, A. W., & Feder, M. A. (Eds.). (2009). Learning science in informal environments: People, places, and pursuits. National Academies Press.
12. Fenichel, M., & Schweingruber, H. A. (2010). Surrounded by science: Learning science in informal environments. National Academies Press.
13. Haden, C. A., Jant, E. A., Hoffman, P. C., Marcus, M., Geddes, J. R., & Gaskins, S. (2014). Supporting family conversations and children's STEM learning in a children's museum. Early Childhood Research Quarterly, 29(3), 333-344.
14. Bevan, B., Gutwill, J. P., Petrich, M., & Wilkinson, K. (2015). Learning through STEM-rich tinkering: Findings from a jointly negotiated research project taken up in practice. Science Education, 99(1), 98-120.
15. Krishnamurthi, A., Ballard, M., & Noam, G. G. (2014). Examining the impact of afterschool STEM programs. Afterschool Alliance.
16. Bonney, R., Cooper, C. B., Dickinson, J., Kelling, S., Phillips, T., Rosenberg, K. V., & Shirk, J. (2009). Citizen science: A developing tool for expanding science knowledge and scientific literacy. BioScience, 59(11), 977-984.
17. Brossard, D., Lewenstein, B., & Bonney, R. (2005). Scientific knowledge and attitude change: The impact of a citizen science project. International Journal of Science Education, 27(9), 1099-1121.

Successful Programs and Initiatives

As the importance of scientific literacy continues to grow in the 21st century, educators, policymakers, and community leaders have developed a range of innovative programs and initiatives to improve science education and engagement. These programs have taken many forms, from school-based interventions and curriculum reforms to informal learning experiences and community partnerships. By examining some of the most successful examples of these programs, we can gain valuable insights into what works in science education and how to scale up these approaches for greater impact.

One highly effective model for improving science education is the use of inquiry-based learning, which engages students in the process of scientific discovery through hands-on, experiential activities [1]. The "Exploring Computer Science" program, developed by the University of Oregon, is a prime example of this approach. This program provides a comprehensive curriculum for high school students that emphasizes problem-solving, creativity, and collaboration through a series of project-based units [2]. By focusing on real-world applications of computer science, such as web design and data analysis, the program has been shown to increase student interest and engagement in the field, particularly among underrepresented groups [3].

Another successful initiative is the "Science Olympiad," a national competition that challenges teams of middle and high school students to demonstrate their knowledge and skills in a variety of scientific disciplines [4]. Through a series of regional and state competitions, students work together to solve problems, design experiments, and build devices related to topics such as astronomy, biology, and engineering. The program has been shown to increase student motivation, self-efficacy, and interest in STEM careers, as well as to promote teamwork and leadership skills [5].

Informal science education programs, such as museums and science centers, have also played a key role in promoting scientific literacy and engagement. The "Exploratorium" in San Francisco, for example, is a pioneering institution that has been at the forefront of hands-on, interactive science education for over 50 years [6]. Through a wide range of exhibits, workshops, and outreach programs, the Exploratorium engages visitors of all ages in the wonder and excitement of scientific discovery. One of its most successful initiatives is the "Tinkering Studio," a space where visitors can design, build, and test their own creations using a variety of tools and materials [7]. This approach has been shown to foster creativity, problem-solving skills, and a deeper understanding of scientific concepts [8].

Community partnerships and outreach programs have also been effective in bringing science education to underserved populations. The "Detroit Area Pre-College Engineering Program" (DAPCEP), for example, is a collaborative effort between universities, schools, and industry partners to provide STEM enrichment opportunities for K-12 students in the Detroit area [9]. Through summer camps, Saturday academies, and after-school programs, DAPCEP exposes students to hands-on engineering and science activities, as well as mentoring and career exploration opportunities. The program has been successful in increasing the number of underrepresented students who pursue STEM degrees and careers [10].

Teacher professional development is another critical component of improving science education. The "National Science Teaching Association" (NSTA) is a leading organization that provides a wide range of resources and support for science teachers at all levels [11]. One of its most successful initiatives is the "New Science Teacher Academy," a year-long program that provides mentoring, online resources, and face-to-face workshops for early-career science teachers [12]. By supporting teachers in their first few years in the classroom, the program has been shown to increase teacher retention, effectiveness, and job satisfaction [13].

Curriculum reform efforts have also played a significant role in improving science education. The "Next Generation Science Standards" (NGSS), developed by a consortium of states and scientific organizations, represent a major shift in the way science is taught in K-12 schools [14]. The standards emphasize the integration of scientific practices, crosscutting concepts, and disciplinary core ideas, as well as the use of performance expectations to assess student learning [15]. While the implementation of the NGSS has been challenging in some contexts, early research suggests that they have the potential to improve student engagement, understanding, and achievement in science [16].

Finally, citizen science programs have emerged as a promising way to engage the public in authentic scientific research and to promote scientific literacy and environmental awareness. The "Galaxy Zoo" project, for example, invites volunteers from around the world to classify images of galaxies captured by the Hubble Space Telescope [17]. By harnessing the power of crowdsourcing, the project has been able to analyze millions of images and to make significant contributions to the field of astronomy [18]. Similarly, the "Monarch Larvae Monitoring Project" engages volunteers in collecting data on monarch butterfly populations across North America, providing valuable information for conservation efforts [19].

These examples represent just a small sample of the many successful programs and initiatives that are working to improve science education and engagement around the world. While each program is unique in its approach and context, they share some common elements that contribute to their effectiveness. These include a focus on hands-on, experiential learning; the use of real-world applications and problems; the engagement of diverse learners and communities; and the cultivation of partnerships and collaborations across sectors and disciplines.

As we look to the future of science education, it is clear that we will need to continue to innovate and experiment with new approaches and models. We will need to harness the power of technology and data to personalize learning and to support teachers in their work. We will need to address issues of equity and access, to ensure that all students have the opportunity to engage in high-quality science education. And we will need to build stronger connections between formal and informal learning environments, between schools and communities, and between science and society as a whole.

By learning from the successes of programs like those described here, and by continuing to push the boundaries of what is possible in science education, we can create a future in which scientific literacy is a fundamental right and a shared value for all. We can empower a new generation of scientists, innovators, and informed citizens who are equipped to tackle the complex challenges of the 21st century, from climate change and public health to space exploration and artificial intelligence. And we can build a society that values science not just as a means to an end, but as a way of understanding ourselves and our place in the universe.

References

1. Pedaste, M., Mäeots, M., Siiman, L. A., De Jong, T., Van Riesen, S. A., Kamp, E. T., ... & Tsourlidaki, E. (2015). Phases of inquiry-based learning: Definitions and the inquiry cycle. Educational Research Review, 14, 47-61.
2. Goode, J., & Margolis, J. (2011). Exploring computer science: A case study of school reform. ACM Transactions on Computing Education (TOCE), 11(2), 1-16.
3. Margolis, J., Estrella, R., Goode, J., Holme, J. J., & Nao, K. (2017). Stuck in the shallow end: Education, race, and computing. MIT Press.
4. Wirt, J. L. (2011). An analysis of Science Olympiad participants' perceptions regarding their experience with the science and engineering academic competition. Dissertations, 26.
5. Sahin, A. (2013). STEM clubs and science fair competitions: Effects on post-secondary matriculation. Journal of STEM Education: Innovations and Research, 14(1), 5.
6. Cole, K. C. (2009). Something incredibly wonderful happens: Frank Oppenheimer and the world he made up. HMH.

7. Petrich, M., Wilkinson, K., & Bevan, B. (2013). It looks like fun, but are they learning?. In Design, make, play (pp. 68-88). Routledge.
8. Bevan, B., Gutwill, J. P., Petrich, M., & Wilkinson, K. (2015). Learning through STEM-rich tinkering: Findings from a jointly negotiated research project taken up in practice. Science Education, 99(1), 98-120.
9. Volin, K. J. (2014). The Detroit Area Pre-College Engineering Program: A descriptive study of the pre-college STEM program and its influence on student post-secondary enrollments (Doctoral dissertation, Wayne State University).
10. Davis, C. E., & Yeary, M. B. (2012). The DAPCEP experience with underserved youth: Parental involvement, STEM education, and testing. In 2012 IEEE International Symposium on Antennas and Propagation (pp. 1-2). IEEE.
11. National Science Teaching Association. (2020). About NSTA. https://www.nsta.org/about
12. Teed, R., Franco, S., & Kind, J. (2013). The new science teacher academy: Mentor involvement, perceived benefits, and mentor needs. Journal of Geoscience Education, 61(4), 405-412.
13. Kelly, M. J., Randle, L., & Barlow, M. (2013). New science teacher academy: building a community of practice to support beginning teachers. In Exemplary Science: Best Practices in Professional Development, Revised Second Edition (pp. 199-211). NSTA Press.
14. NGSS Lead States. (2013). Next generation science standards: For states, by states. Washington, DC: The National Academies Press.
15. Pruitt, S. L. (2014). The next generation science standards: The features and challenges. Journal of Science Teacher Education, 25(2), 145-156.
16. Harris, K., Sithole, A., & Kibirige, J. (2017). A needs assessment for the adoption of Next Generation Science Standards (NGSS) in K-12 education in the United States. Journal of Education and Training Studies, 5(9), 54-62.
17. Raddick, M. J., Bracey, G., Gay, P. L., Lintott, C. J., Murray, P., Schawinski, K., ... & Vandenberg, J. (2009). Galaxy zoo: Exploring the motivations of citizen science volunteers. arXiv preprint arXiv:0909.2925.
18. Cardamone, C., Schawinski, K., Sarzi, M., Bamford, S. P., Bennert, N., Urry, C. M., ... & VandenBerg, J. (2009). Galaxy Zoo Green Peas: Discovery of a class of compact extremely star-forming galaxies. Monthly Notices of the Royal Astronomical Society, 399(3), 1191-1205.
19. Ries, L., Taron, D. J., & Rendón-Salinas, E. (2015). The disconnect between summer and winter monarch trends for the eastern migratory population: Possible links to differing drivers. Annals of the Entomological Society of America, 108(5), 691-699.

Chapter 9: Promoting Science Communication and Public Engagement

The Importance of Effective Science Communication

In an era of rapid scientific and technological advancement, the ability to communicate science effectively to a wide range of audiences has become increasingly critical. Science communication encompasses a variety of practices and strategies aimed at informing, educating, and engaging the public in scientific topics and issues [1]. From helping individuals make informed decisions about their health and well-being to shaping public policy and inspiring the next generation of scientists, effective science communication has the potential to transform society in profound ways.

At its core, science communication is about bridging the gap between the scientific community and the general public. Scientists and researchers have a wealth of knowledge and expertise that can benefit society in countless ways, but this knowledge is often complex, technical, and difficult for non-experts to understand [2]. Science communicators play a crucial role in translating scientific findings and concepts into language and formats that are accessible, engaging, and relevant to diverse audiences [3].

One of the primary goals of science communication is to promote scientific literacy and understanding among the general public. By providing accurate, up-to-date information about scientific topics and issues, science communicators can help individuals develop a better understanding of the world around them and make more informed decisions in their personal and professional lives [4]. This is particularly important in areas such as health and medicine, where a lack of scientific understanding can lead to poor health choices, mistrust of medical professionals, and the spread of misinformation [5].

Effective science communication can also play a key role in shaping public policy and decision-making. Many of the most pressing challenges facing society today, from climate change and infectious diseases to food security and energy production, require a deep understanding of scientific principles and evidence [6]. By communicating the latest scientific findings and their implications to policymakers and other stakeholders, science communicators can help ensure that decisions are based on sound science rather than political ideology or special interests [7].

In addition to informing and educating the public, science communication can also serve to inspire and engage individuals in scientific pursuits. By sharing the excitement and wonder of scientific discovery, science communicators can help spark curiosity and interest in science, particularly among young people [8]. This is crucial for developing the next generation of scientists, engineers, and innovators who will drive scientific and technological progress in the years to come [9].

However, effective science communication is not without its challenges. One of the biggest obstacles is the complexity and uncertainty inherent in much of scientific research. Unlike other fields where facts and figures can be easily distilled into sound bites or infographics, science often involves nuance, caveats, and ongoing debates [10]. Science communicators must find ways to convey this complexity and uncertainty without oversimplifying or misrepresenting the science, while also making it accessible and engaging to non-expert audiences [11].

Another challenge is the proliferation of misinformation and pseudoscience in the digital age. With the rise of social media and other online platforms, it has become easier than ever for false or misleading information to spread rapidly and reach large audiences [12]. Science communicators must contend with a constant stream of competing messages and narratives, some of which are deliberately designed to undermine public trust in science and evidence-based decision-making [13].

To overcome these challenges, science communicators must employ a range of strategies and best practices. One key approach is to use storytelling and narrative techniques to make scientific information more engaging and memorable [14]. By weaving scientific facts and concepts into compelling stories and examples, science communicators can help audiences connect with the science on a personal and emotional level, making it more likely to stick in their minds and influence their attitudes and behaviors [15].

Another important strategy is to tailor science communication to the needs and interests of specific audiences. Different groups of people have different levels of scientific knowledge, different communication preferences, and different stakes in scientific issues [16]. By understanding these differences and adapting their communication strategies accordingly, science communicators can more effectively reach and engage their target audiences [17].

Effective science communication also requires building trust and credibility with audiences. This means being transparent about the sources and methods of scientific research, acknowledging uncertainties and limitations, and engaging in open and honest dialogue with the public [18]. It also means working to diversify the voices and perspectives in science communication, so that it reflects the diversity of the scientific community and the broader society [19].

Ultimately, the importance of effective science communication lies in its ability to empower individuals and society as a whole to make informed decisions and take action on the issues that matter most. By providing accurate, accessible, and engaging information about scientific topics and issues, science communicators can help create a more scientifically literate and engaged public, one that is better equipped to tackle the complex challenges of the 21st century.

As Carl Sagan, one of the most influential science communicators of the 20th century, once said: "We live in a society exquisitely dependent on science and technology, in which hardly anyone knows anything about science and technology." [20] Effective science communication is essential for bridging this gap and ensuring that the benefits of scientific knowledge and innovation are shared by all. It is a vital tool for building a more informed, engaged, and empowered society, one that values science as a key driver of progress and a fundamental part of our shared human experience.

References

1. Burns, T. W., O'Connor, D. J., & Stocklmayer, S. M. (2003). Science communication: a contemporary definition. Public Understanding of Science, 12(2), 183-202.
2. National Academies of Sciences, Engineering, and Medicine. (2017). Communicating science effectively: a research agenda. National Academies Press.
3. Fischhoff, B. (2013). The sciences of science communication. Proceedings of the National Academy of Sciences, 110(Supplement 3), 14033-14039.
4. Nisbet, M. C., & Scheufele, D. A. (2009). What's next for science communication? Promising directions and lingering distractions. American Journal of Botany, 96(10), 1767-1778.
5. Ratzan, S. C. (2001). Health literacy: communication for the public good. Health Promotion International, 16(2), 207-214.
6. Fischhoff, B. (2019). Evaluating science communication. Proceedings of the National Academy of Sciences, 116(16), 7670-7675.
7. Scheufele, D. A. (2014). Science communication as political communication. Proceedings of the National Academy of Sciences, 111(Supplement 4), 13585-13592.
8. Bell, P., Lewenstein, B., Shouse, A. W., & Feder, M. A. (Eds.). (2009). Learning science in informal environments: People, places, and pursuits. National Academies Press.
9. Falk, J. H., & Dierking, L. D. (2010). The 95 percent solution. American Scientist, 98(6), 486-493.
10. Zehr, S. C. (2000). Public representations of scientific uncertainty about global climate change. Public Understanding of Science, 9(2), 85-103.
11. Friedman, S. M., Dunwoody, S., & Rogers, C. L. (Eds.). (1999). Communicating uncertainty: Media coverage of new and controversial science. Routledge.
12. Scheufele, D. A., & Krause, N. M. (2019). Science audiences, misinformation, and fake news. Proceedings of the National Academy of Sciences, 116(16), 7662-7669.
13. Iyengar, S., & Massey, D. S. (2019). Scientific communication in a post-truth society. Proceedings of the National Academy of Sciences, 116(16), 7656-7661.
14. Dahlstrom, M. F. (2014). Using narratives and storytelling to communicate science with nonexpert audiences. Proceedings of the National Academy of Sciences, 111(Supplement 4), 13614-13620.
15. Green, M. C., Brock, T. C., & Kaufman, G. F. (2004). Understanding media enjoyment: The role of transportation into narrative worlds. Communication Theory, 14(4), 311-327.
16. Nisbet, M. C., & Mooney, C. (2007). Framing science. Science, 316(5821), 56.
17. Brossard, D., & Lewenstein, B. V. (2010). A critical appraisal of models of public understanding of science: Using practice to inform theory. In L. Kahlor & P. Stout (Eds.), Communicating science: New agendas in communication (pp. 11-39). Routledge.
18. Resnick, B. (2019). The war on truth is a war on democracy. Vox. https://www.vox.com/policy-and-politics/2019/1/10/18175913/brexit-trump-truth-democracy-psychology-epistemic-crisis

19. Dudo, A. (2015). Scientists, the media, and the public communication of science. Sociology Compass, 9(9), 761-775.
20. Sagan, C. (1990). Why we need to understand science. Skeptical Inquirer, 14(3), 263-269.

Strategies for Scientists, Journalists, and Institutions

Effective science communication is a shared responsibility that requires the active participation and collaboration of scientists, journalists, and institutions. Each of these groups plays a unique and critical role in shaping public understanding and engagement with science, and each faces its own set of challenges and opportunities in the rapidly evolving media landscape of the 21st century. By developing and implementing targeted strategies for science communication, these groups can work together to create a more informed, engaged, and scientifically literate society.

For scientists, effective communication is no longer an optional extra but an essential part of their professional responsibilities [1]. As the primary producers of scientific knowledge, scientists have a unique obligation to share their findings and expertise with the public in ways that are accurate, accessible, and engaging [2]. This means going beyond simply publishing their work in academic journals and presenting at conferences, and actively seeking out opportunities to communicate with broader audiences through media interviews, public lectures, social media, and other outreach activities [3].

To be effective communicators, scientists need to develop a range of skills and strategies that go beyond their traditional training. This includes learning how to distill complex scientific concepts into clear, jargon-free language that can be understood by non-experts, and how to use analogies, metaphors, and storytelling techniques to make their work more relatable and memorable [4]. It also means understanding the needs and interests of different audiences, and tailoring their communication strategies accordingly [5].

One promising approach for scientists is to collaborate with communication professionals, such as journalists, public information officers, and science writers, who can help them translate their work for wider audiences [6]. By working with these intermediaries, scientists can benefit from their expertise in crafting compelling narratives, navigating media landscapes, and engaging with diverse publics [7]. However, such collaborations also require scientists to relinquish some control over the communication process and to trust in the skills and integrity of their partners [8].

For journalists, covering science in an accurate, nuanced, and engaging way is a daunting challenge. Science is a complex, dynamic, and often uncertain enterprise, with multiple stakeholders, competing interests, and evolving norms and practices [9]. Journalists must navigate this terrain while also grappling with the economic and technological disruptions that have transformed the news industry in recent years, from the decline of traditional business models to the rise of social media and fake news [10].

To meet these challenges, journalists need to develop a deep understanding of the scientific process, the culture of science, and the social and political contexts in which science operates [11]. This means cultivating sources and relationships with scientists, staying abreast of the latest research and debates in their fields, and learning how to evaluate the credibility and significance of scientific claims [12]. It also means being transparent about the limitations and uncertainties of science, and resisting the temptation to sensationalize or oversimplify complex issues for the sake of a good story [13].

One strategy that has shown promise in recent years is the development of specialized science journalism training programs and fellowships, which provide journalists with the knowledge, skills, and networks they need to cover science effectively [14]. These programs, offered by universities, research institutions, and professional organizations, can help bridge the gap between scientists and journalists and foster greater mutual understanding and respect [15].

Institutions, such as universities, research organizations, and government agencies, also have a critical role to play in promoting effective science communication. As the primary funders and conductors of scientific research, these institutions have a responsibility to ensure that their work is communicated accurately and effectively to the public, and to support and incentivize their researchers to engage in outreach and communication activities [16].

One way institutions can do this is by establishing dedicated science communication offices or teams, which can provide training, resources, and support for researchers looking to engage with the public [17]. These offices can also serve as a central hub for media relations, social media, and other communication activities, helping to ensure that the institution's research is disseminated widely and strategically [18].

Another strategy is to integrate science communication training into graduate and postdoctoral education, so that the next generation of scientists is equipped with the skills and knowledge they need to engage with the public from the outset of their careers [19]. This can include coursework in communication theory and practice, as well as opportunities for hands-on experience through internships, outreach programs, and media placements [20].

Institutions can also play a key role in fostering a culture of science communication and public engagement, by recognizing and rewarding researchers who excel in these areas [21]. This can involve everything from including communication activities in tenure and promotion criteria to providing funding and resources for outreach projects and events [22].

Ultimately, promoting effective science communication requires a sustained and collaborative effort from all stakeholders in the scientific enterprise. By working together to develop and implement strategies that prioritize accuracy, engagement, and accessibility, scientists, journalists, and institutions can help create a more informed and empowered public that is better equipped to make decisions about the complex scientific and technological issues that shape our world.

As the pace of scientific discovery and technological change continues to accelerate, the need for effective science communication has never been greater. From climate change and public health to artificial intelligence and space exploration, the challenges and opportunities facing society in the 21st century will require a deep understanding of science and a willingness to engage with scientific ideas and evidence. By embracing the vital role of science communication and working to enhance its impact and reach, we can help ensure that science continues to serve as a powerful tool for understanding our world and improving our lives.

References

1. Baron, N. (2010). Escape from the ivory tower: A guide to making your science matter. Island Press.
2. Brownell, S. E., Price, J. V., & Steinman, L. (2013). Science communication to the general public: Why we need to teach undergraduate and graduate students this skill as part of their formal scientific training. Journal of Undergraduate Neuroscience Education, 12(1), E6-E10.
3. Brossard, D., & Scheufele, D. A. (2013). Science, new media, and the public. Science, 339(6115), 40-41.
4. Olson, R. (2009). Don't be such a scientist: Talking substance in an age of style. Island Press.
5. Nisbet, M. C., & Scheufele, D. A. (2009). What's next for science communication? Promising directions and lingering distractions. American Journal of Botany, 96(10), 1767-1778.
6. Peters, H. P. (2013). Gap between science and media revisited: Scientists as public communicators. Proceedings of the National Academy of Sciences, 110(Supplement 3), 14102-14109.
7. Dunwoody, S. (2008). Science journalism. In M. Bucchi & B. Trench (Eds.), Handbook of public communication of science and technology (pp. 15-26). Routledge.
8. Rödder, S. (2012). The ambivalence of visible scientists. In S. Rödder, M. Franzen, & P. Weingart (Eds.), The sciences' media connection – Public communication and its repercussions (pp. 155-177). Springer.
9. Friedman, S. M., Dunwoody, S., & Rogers, C. L. (Eds.). (1999). Communicating uncertainty: Media coverage of new and controversial science. Routledge.
10. Schäfer, M. S. (2017). How changing media structures are affecting science news coverage. In K. H. Jamieson, D. M. Kahan, & D. A. Scheufele (Eds.), The Oxford handbook of the science of science communication (pp. 51-59). Oxford University Press.
11. Blum, D., Knudson, M., & Henig, R. M. (Eds.). (2006). A field guide for science writers. Oxford University Press.
12. Hayden, T., & Hayden, E. C. (2018). Science journalism's unlikely golden age. Frontiers in Communication, 2, 24.
13. Stocking, S. H. (1999). How journalists deal with scientific uncertainty. In S. M. Friedman, S. Dunwoody, & C. L. Rogers (Eds.), Communicating uncertainty: Media coverage of new and controversial science (pp. 23-42). Routledge.
14. National Academies of Sciences, Engineering, and Medicine. (2017). Communicating science effectively: A research agenda. National Academies Press.
15. Besley, J. C., & Tanner, A. H. (2011). What science communication scholars think about training scientists to communicate. Science Communication, 33(2), 239-263.
16. Dudo, A., & Besley, J. C. (2016). Scientists' prioritization of communication objectives for public engagement. PLOS ONE, 11(2), e0148867.
17. Smith, B., Baron, N., English, C., Galindo, H., Goldman, E., McLeod, K., Miner, M., & Neeley, E. (2013). COMPASS: Navigating the rules of scientific engagement. PLOS Biology, 11(4), e1001552.

18. Marcinkowski, F., Kohring, M., Fürst, S., & Friedrichsmeier, A. (2014). Organizational influence on scientists' efforts to go public: An empirical investigation. Science Communication, 36(1), 56-80.
19. Baram-Tsabari, A., & Lewenstein, B. V. (2017). Science communication training: What are we trying to teach? International Journal of Science Education, Part B, 7(3), 285-300.
20. Kahan, D. M. (2015). What is the "science of science communication"? Journal of Science Communication, 14(3), Y04.
21. Jacobson, N., Butterill, D., & Goering, P. (2004). Organizational factors that influence university-based researchers' engagement in knowledge transfer activities. Science Communication, 25(3), 246-259.
22. Whitmer, A., Ogden, L., Lawton, J., Sturner, P., Groffman, P. M., Schneider, L., Hart, D., Halpern, B., Schlesinger, W., Raciti, S., Bettez, N., Ortega, S., Rustad, L., Pickett, S. T. A., & Killilea, M. (2010). The engaged university: Providing a platform for research that transforms society. Frontiers in Ecology and the Environment, 8(6), 314-321.

Successful Examples of Public Engagement with Science

Engaging the public with science is a crucial aspect of promoting scientific literacy, fostering trust in scientific institutions, and ensuring that scientific knowledge is used to inform decision-making at all levels of society. In recent years, there have been numerous successful examples of public engagement initiatives that have demonstrated the power of science communication to inspire, educate, and empower citizens to engage with scientific issues that affect their lives and communities.

One notable example is the "Citizen Science" movement, which involves the participation of non-professional scientists in scientific research projects [1]. Citizen science initiatives have been used to engage the public in a wide range of scientific fields, from astronomy and ecology to public health and climate change [2]. By inviting members of the public to collect data, analyze results, and contribute their knowledge and expertise, citizen science projects can help to democratize science and make it more accessible and relevant to people's everyday lives [3].

One of the most successful citizen science projects to date is "Galaxy Zoo," an online platform that invites users to classify images of galaxies captured by telescopes [4]. Since its launch in 2007, Galaxy Zoo has engaged more than 150,000 volunteers from around the world, who have collectively classified more than 50 million galaxies [5]. The project has led to numerous scientific discoveries, including the identification of rare types of galaxies and the mapping of the large-scale structure of the universe [6]. It has also inspired a host of other citizen science projects in fields ranging from marine biology to neuroscience [7].

Another successful example of public engagement with science is the "Science Café" movement, which involves informal gatherings in public spaces where scientists and members of the public can engage in conversations about scientific topics [8]. Science cafés originated in the UK in the late 1990s and have since spread to more than 40 countries around the world [9]. These events typically involve a scientist giving a short presentation on a topic of interest, followed by an open discussion with the audience [10].

Science cafés have been used to engage the public in a wide range of scientific issues, from the science of cooking and the psychology of happiness to the ethics of gene editing and the search for extraterrestrial life [11]. By creating a relaxed and informal atmosphere, science cafés can help to break down barriers between scientists and the public and foster a sense of shared curiosity and discovery [12]. They can also provide a platform for scientists to practice their communication skills and gain valuable feedback from a diverse audience [13].

Another innovative approach to public engagement with science is the use of "Science Festivals," which are multi-day events that celebrate science through a variety of interactive exhibits, demonstrations, and performances [14]. Science festivals have become increasingly popular in recent years, with more than 500 events taking place around the world each year [15]. These events typically involve collaborations between universities, museums, schools, and community organizations, and attract audiences ranging from young children to senior citizens [16].

One of the most successful science festivals is the "Edinburgh International Science Festival," which has been running annually since 1989 [17]. The festival attracts more than 150,000 visitors each year and features more than 200 events, including talks, workshops, exhibitions, and hands-on activities [18]. The festival has a particular focus on engaging young people with science, with a dedicated program for schools and families [19]. It has also been used as a model for other science festivals around the world, including the "USA Science & Engineering Festival" in Washington, DC, and the "World Science Festival" in New York City [20].

A fourth example of successful public engagement with science is the use of "Science Museums," which are institutions dedicated to showcasing scientific discoveries and promoting scientific literacy [21]. Science museums have a long history dating back to the 17th century, but in recent years they have become increasingly interactive and focused on engaging visitors with hands-on exhibits and activities [22].

One of the most successful science museums is the "Exploratorium" in San Francisco, which was founded in 1969 and has since become a model for interactive science education around the world [23]. The Exploratorium features more than 600 exhibits on topics ranging from human perception to the physics of light and sound [24]. It also offers a wide range of educational programs for students, teachers, and the general public, including workshops, lectures, and online resources [25].

Science museums have been shown to have a significant impact on public attitudes towards science and technology. A study of visitors to the "Museum of Science and Industry" in Chicago found that the museum experience increased visitors' interest in science, their understanding of scientific concepts, and their appreciation of the relevance of science to their daily lives [26]. Another study of the "London Science Museum" found that the museum's exhibits and programs helped to challenge stereotypes about science and scientists and promote a more inclusive vision of science [27].

These examples illustrate the diverse range of approaches that can be used to engage the public with science, from citizen science projects and science cafés to science festivals and museums. What all of these initiatives have in common is a commitment to making science accessible, relevant, and engaging to a wide audience, and to fostering a sense of shared ownership and participation in the scientific enterprise.

As we look to the future, it is clear that public engagement with science will become increasingly important in addressing the complex challenges facing our world, from climate change and public health to food security and technological disruption. By continuing to develop and support innovative approaches to science communication and engagement, we can help to build a more scientifically literate and engaged society, one that is better equipped to make informed decisions and take action on the issues that matter most.

Ultimately, the success of public engagement with science will depend on the active participation and collaboration of scientists, educators, policymakers, and citizens alike. By working together to promote a culture of scientific curiosity, creativity, and critical thinking, we can help to ensure that science remains a vital and transformative force for good in the world, and that its benefits are shared by all.

References

1. Bonney, R., Shirk, J. L., Phillips, T. B., Wiggins, A., Ballard, H. L., Miller-Rushing, A. J., & Parrish, J. K. (2014). Next steps for citizen science. Science, 343(6178), 1436-1437.
2. Dickinson, J. L., & Bonney, R. (Eds.). (2012). Citizen science: Public participation in environmental research. Cornell University Press.
3. Irwin, A. (1995). Citizen science: A study of people, expertise and sustainable development. Routledge.
4. Raddick, M. J., Bracey, G., Gay, P. L., Lintott, C. J., Murray, P., Schawinski, K., ... & Vandenberg, J. (2009). Galaxy Zoo: Exploring the motivations of citizen science volunteers. arXiv preprint arXiv:0909.2925.
5. Lintott, C., Schawinski, K., Bamford, S., Slosar, A., Land, K., Thomas, D., ... & Vandenberg, J. (2010). Galaxy Zoo 1: Data release of morphological classifications for nearly 900,000 galaxies. Monthly Notices of the Royal Astronomical Society, 410(1), 166-178.
6. Cardamone, C., Schawinski, K., Sarzi, M., Bamford, S. P., Bennert, N., Urry, C. M., ... & VandenBerg, J. (2009). Galaxy Zoo Green Peas: Discovery of a class of compact extremely star-forming galaxies. Monthly Notices of the Royal Astronomical Society, 399(3), 1191-1205.
7. Bonney, R., Phillips, T. B., Ballard, H. L., & Enck, J. W. (2016). Can citizen science enhance public understanding of science? Public Understanding of Science, 25(1), 2-16.
8. Dallas, D. (1999). The Café Scientifique. Nature, 399(6732), 120-120.
9. Giles, J. (2004). Pop science pulls in public. Nature, 429(6987), 7-7.
10. Cohen, J., & Macfarlane, H. (2007). Beer and bosons at the Café Scientifique. Museums & Social Issues, 2(2), 233-242.
11. Grand, A. (2014). Café Scientifique. Science Progress, 97(3), 275-278.
12. Navid, E. L., & Einsiedel, E. F. (2012). Synthetic biology in the Science Café: What have we learned about public engagement? Journal of Science Communication, 11(4), A02.
13. Dijkstra, A. M. (2017). Analysing Dutch Science Cafés to better understand the science-society relationship. Journal of Science Communication, 16(1), A03.

14. Bultitude, K. (2014). Science festivals: Do they succeed in reaching beyond the "already engaged"? Journal of Science Communication, 13(4), C01.
15. Wiehe, B. (2014). When science makes us who we are: Known and speculative impacts of science festivals. Journal of Science Communication, 13(4), C02.
16. Fogg-Rogers, L., Bay, J. L., Burgess, H., & Purdy, S. C. (2015). 'Knowledge is power': A mixed-methods study exploring adult audience preferences for engagement and learning formats over 3 years of a health science festival. Science Communication, 37(4), 419-451.
17. Gage, S. (2001). Edinburgh International Science Festival. In Science Communication in Theory and Practice (pp. 203-217). Springer.
18. Edinburgh International Science Festival. (2020). About the Festival. https://www.sciencefestival.co.uk/about-the-festival
19. Jensen, E., & Buckley, N. (2014). Why people attend science festivals: Interests, motivations and self-reported benefits of public engagement with research. Public Understanding of Science, 23(5), 557-573.
20. Rose, K. M., Korzekwa, K., Brossard, D., Scheufele, D. A., & Heisler, L. (2017). Engaging the public at a science festival: Findings from a panel on human gene editing. Science Communication, 39(2), 250-277.
21. Friedman, A. J. (2010). The evolution of the science museum. Physics Today, 63(10), 45-51.
22. Schiele, B. (2014). Science museums and science centres. In Routledge Handbook of Public Communication of Science and Technology (pp. 40-53). Routledge.
23. Oppenheimer, F. (1972). The Exploratorium: A playful museum combines perception and art in science education. American Journal of Physics, 40(7), 978-984.
24. Exploratorium. (2020). Exhibits. https://www.exploratorium.edu/exhibits
25. Exploratorium. (2020). Education. https://www.exploratorium.edu/education
26. Falk, J. H., & Needham, M. D. (2011). Measuring the impact of a science center on its community. Journal of Research in Science Teaching, 48(1), 1-12.
27. Tlili, A., Cribb, A., & Gewirtz, S. (2006). What becomes of science in a science centre? Reconfiguring science for public consumption at the UK's Science Museum. Review of Education, Pedagogy, and Cultural Studies, 28(2), 203-228.

Chapter 10: Building a Scientifically Literate Society

The Benefits of a Scientifically Literate Population

The importance of scientific literacy in today's rapidly changing world cannot be overstated. As our society becomes increasingly complex and technology-driven, the ability to understand and apply scientific principles and methods is essential for personal, professional, and civic success. A scientifically literate population is better equipped to make informed decisions, engage in critical thinking, and contribute to the economic and social well-being of their communities and the world at large.

One of the most significant benefits of a scientifically literate population is the ability to make informed decisions about personal health and well-being. With the proliferation of health information available online and through various media channels, individuals must be able to distinguish between credible, evidence-based advice and misinformation or pseudoscience. Scientific literacy enables individuals to understand the basic principles of human biology, nutrition, and disease prevention, and to evaluate the risks and benefits of different medical treatments or health behaviors [1]. This knowledge can lead to better health outcomes, reduced healthcare costs, and improved quality of life [2].

Moreover, scientific literacy is crucial for making informed decisions about environmental and social issues that affect our planet and our communities. Climate change, for example, is one of the most pressing challenges facing humanity today, with far-reaching consequences for our economy, health, and way of life [3]. A scientifically literate population is better equipped to understand the evidence behind climate change, to evaluate the potential impacts of different policy options, and to take individual and collective action to mitigate its effects [4]. Similarly, scientific literacy can help individuals navigate complex social issues such as vaccine hesitancy, genetic engineering, and artificial intelligence, and to participate in informed public discourse about the ethical and societal implications of these technologies [5].

In addition to personal and societal benefits, scientific literacy is also essential for economic growth and competitiveness in the 21st century. As the global economy becomes increasingly knowledge-based and innovation-driven, the demand for workers with strong scientific and technical skills continues to grow [6]. A scientifically literate workforce is better prepared to adapt to new technologies, to innovate and create new products and services, and to compete in the global marketplace [7]. Moreover, a population that values and supports scientific research and development is more likely to invest in and benefit from the economic and social returns of these investments [8].

Scientific literacy is also essential for effective citizenship and democratic participation. In a world where science and technology are increasingly intertwined with public policy and decision-making, individuals must be able to evaluate the evidence behind different policy options, to understand the potential risks and benefits of different courses of action, and to hold their elected officials accountable for making informed and evidence-based decisions [9]. A scientifically literate population is more likely to support policies that are based on sound scientific evidence and to reject policies that are based on misinformation or ideology [10].

Furthermore, scientific literacy can promote greater public trust and confidence in science and scientific institutions. When individuals understand the basic principles and methods of science, they are more likely to appreciate the value and limitations of scientific knowledge, and to trust the findings of scientific research even when they may be counterintuitive or controversial [11]. This trust is essential for the effective functioning of our scientific institutions and for the continued advancement of scientific knowledge [12]. In contrast, a lack of scientific literacy can lead to mistrust, skepticism, and even hostility towards science and scientists, as evidenced by recent public debates over issues such as climate change, vaccination, and GMOs [13].

Finally, scientific literacy can enhance individuals' intellectual curiosity, creativity, and sense of wonder about the natural world. By understanding the basic principles that govern the universe, from the smallest subatomic particles to the largest galaxies, individuals can gain a deeper appreciation for the beauty and complexity of the world around them [14]. This sense of wonder and curiosity can inspire individuals to pursue careers in science and technology, to engage in lifelong learning and discovery, and to contribute to the advancement of human knowledge and understanding [15].

To realize these benefits, however, it is essential to cultivate scientific literacy throughout all stages of life and across all segments of society. This requires a comprehensive and multifaceted approach that includes formal education in schools and universities, informal learning opportunities through museums, media, and community programs, and public engagement and outreach by scientists and scientific institutions [16]. It also requires addressing issues of equity and inclusion in science education and careers, to ensure that all individuals have access to the knowledge and skills needed to fully participate in the scientific enterprise [17].

Moreover, building a scientifically literate society requires a cultural shift towards valuing and prioritizing science and evidence-based decision-making. This means not only investing in scientific research and education, but also promoting a culture of critical thinking, open-mindedness, and intellectual humility [18]. It means encouraging individuals to ask questions, to seek out reliable sources of information, and to engage in respectful and informed dialogue about scientific issues [19].

Ultimately, the benefits of a scientifically literate population extend far beyond the individual level. A society that values and applies scientific knowledge and methods is better equipped to tackle the complex challenges of the 21st century, from public health and environmental sustainability to economic prosperity and social justice. By investing in scientific literacy and building a culture of evidence-based decision-making, we can create a more informed, engaged, and empowered citizenry, and ensure a brighter and more prosperous future for all.

References

1. Nutbeam, D. (2008). The evolving concept of health literacy. Social Science & Medicine, 67(12), 2072-2078.
2. Berkman, N. D., Sheridan, S. L., Donahue, K. E., Halpern, D. J., & Crotty, K. (2011). Low health literacy and health outcomes: An updated systematic review. Annals of Internal Medicine, 155(2), 97-107.
3. Intergovernmental Panel on Climate Change (IPCC). (2018). Global Warming of 1.5°C: An IPCC Special Report on the impacts of global warming of 1.5°C above pre-industrial levels and related global greenhouse gas emission pathways. IPCC.
4. Hornsey, M. J., & Fielding, K. S. (2017). Attitude roots and Jiu Jitsu persuasion: Understanding and overcoming the motivated rejection of science. American Psychologist, 72(5), 459-473.
5. National Academies of Sciences, Engineering, and Medicine (NASEM). (2016). Science literacy: Concepts, contexts, and consequences. National Academies Press.
6. National Science Board (NSB). (2019). The skilled technical workforce: Crafting America's science & engineering enterprise. NSB.
7. Carnevale, A. P., Smith, N., & Melton, M. (2011). STEM: Science, technology, engineering, mathematics. Georgetown University Center on Education and the Workforce.
8. American Association for the Advancement of Science (AAAS). (2018). Why public engagement matters. AAAS.
9. Miller, J. D. (2016). Civic scientific literacy in the United States in 2016. University of Michigan.
10. Maibach, E., Leiserowitz, A., Roser-Renouf, C., Myers, T., Rosenthal, S., & Feinberg, G. (2015). The Francis effect: How Pope Francis changed the conversation about global warming. George Mason University Center for Climate Change Communication.
11. Drummond, C., & Fischhoff, B. (2017). Individuals with greater science literacy and education have more polarized beliefs on controversial science topics. Proceedings of the National Academy of Sciences, 114(36), 9587-9592.
12. Fiske, S. T., & Dupree, C. (2014). Gaining trust as well as respect in communicating to motivated audiences about science topics. Proceedings of the National Academy of Sciences, 111(Supplement 4), 13593-13597.
13. Bauer, M. W. (2009). The evolution of public understanding of science—discourse and comparative evidence. Science, Technology and Society, 14(3), 221-240.
14. Keltner, D., & Haidt, J. (2003). Approaching awe, a moral, spiritual, and aesthetic emotion. Cognition and Emotion, 17(2), 297-314.
15. Valdesolo, P., Park, J., & Gottlieb, S. (2016). Awe and scientific explanation. Emotion, 16(7), 937-940.
16. Bell, P., Lewenstein, B., Shouse, A. W., & Feder, M. A. (Eds.). (2009). Learning science in informal environments: People, places, and pursuits. National Academies Press.
17. Lee, O., & Buxton, C. A. (2010). Diversity and equity in science education: Research, policy, and practice. Teachers College Press.
18. Sinatra, G. M., Kienhues, D., & Hofer, B. K. (2014). Addressing challenges to public understanding of science: Epistemic cognition, motivated reasoning, and conceptual change. Educational Psychologist, 49(2), 123-138.

19. Dudo, A., & Besley, J. C. (2016). Scientists' prioritization of communication objectives for public engagement. PLoS ONE, 11(2), e0148867.

The Role of Individuals, Communities, and Policymakers

Building a scientifically literate society is a complex and multifaceted endeavor that requires the active participation and collaboration of individuals, communities, and policymakers. Each of these groups has a unique and essential role to play in promoting the value of science, fostering a culture of curiosity and critical thinking, and ensuring that scientific knowledge is applied to inform decision-making at all levels of society.

At the individual level, every person has the power to contribute to a scientifically literate society through their own attitudes, behaviors, and choices. One of the most fundamental ways that individuals can promote scientific literacy is by cultivating a personal interest in and appreciation for science. This means seeking out opportunities to learn about scientific discoveries, engaging with science-related media and events, and sharing one's enthusiasm for science with others [1]. By modeling a positive and curious attitude towards science, individuals can help to create a culture that values and celebrates scientific knowledge and inquiry.

Moreover, individuals can contribute to scientific literacy by being critical consumers of information and by making evidence-based decisions in their personal and professional lives. In an era of "fake news" and misinformation, it is essential for individuals to develop the skills to evaluate the credibility and reliability of different sources of information, and to base their beliefs and actions on sound scientific evidence rather than anecdotes or opinions [2]. This means asking questions, seeking out multiple perspectives, and being willing to change one's mind in light of new evidence [3].

At the community level, there are many ways that groups of individuals can work together to promote scientific literacy and engagement. One important approach is through the development of community-based science programs and initiatives that provide opportunities for people of all ages and backgrounds to participate in scientific research and learning. These programs can take many forms, from citizen science projects that engage volunteers in data collection and analysis, to science festivals and events that showcase the excitement and relevance of science to everyday life [4].

Community-based science programs can also play a key role in addressing issues of equity and inclusion in science education and participation. By partnering with schools, libraries, museums, and other community organizations, these programs can help to reach underserved and underrepresented groups who may not have access to traditional science learning opportunities [5]. Moreover, by involving community members in the design and implementation of these programs, they can help to ensure that the science being conducted is relevant and responsive to the needs and interests of the community [6].

Another way that communities can promote scientific literacy is through the development of local and regional science communication networks and partnerships. These networks can bring together scientists, educators, journalists, and other stakeholders to share knowledge, resources, and best practices for engaging the public with science [7]. By fostering collaboration and dialogue across different sectors and disciplines, these networks can help to break down silos and create a more integrated and effective approach to science communication and outreach.

At the policy level, there are many ways that governments and institutions can support the development of a scientifically literate society. One of the most important roles of policymakers is to ensure that science education is a priority at all levels of the education system, from primary and secondary schools to colleges and universities. This means providing adequate funding and resources for science education programs, as well as setting standards and guidelines for science curriculum and instruction [8].

Policymakers can also support scientific literacy by investing in research and development that addresses societal needs and challenges. By prioritizing funding for research in areas such as public health, environmental sustainability, and technological innovation, policymakers can help to ensure that scientific knowledge is being applied to solve real-world problems and improve people's lives [9]. Moreover, by involving the public in the research process through citizen science and other participatory approaches, policymakers can help to build trust and engagement between the scientific community and the broader society [10].

Another key role of policymakers is to promote the use of scientific evidence in decision-making and policy development. This means creating mechanisms and processes for integrating scientific knowledge into policy discussions and ensuring that decisions are based on the best available evidence rather than political or ideological considerations [11]. It also means being transparent about the scientific basis for policy decisions and communicating the rationale and evidence behind these decisions to the public in clear and accessible ways [12].

Finally, policymakers have a responsibility to address issues of equity and inclusion in science and to ensure that the benefits of scientific knowledge and innovation are shared widely across society. This means investing in programs and initiatives that support the participation and advancement of underrepresented groups in science, as well as ensuring that the products and applications of science are accessible and affordable to all [13].

Ultimately, building a scientifically literate society is a shared responsibility that requires the engagement and collaboration of individuals, communities, and policymakers. By working together to promote the value of science, to foster a culture of curiosity and critical thinking, and to ensure that scientific knowledge is applied to inform decision-making and improve people's lives, we can create a more informed, engaged, and empowered society that is better equipped to tackle the challenges of the 21st century.

This is not an easy task, and it will require sustained effort and investment over time. But the benefits of a scientifically literate society are clear and compelling, from improved health and well-being to enhanced economic competitiveness and social progress. By embracing the role of science in our lives and communities, and by working together to build a culture that values and supports scientific inquiry and knowledge, we can create a brighter and more prosperous future for all.

References

1. Falk, J. H., & Dierking, L. D. (2010). The 95 percent solution: School is not where most Americans learn most of their science. American Scientist, 98(6), 486–493.
2. Scheufele, D. A., & Krause, N. M. (2019). Science audiences, misinformation, and fake news. Proceedings of the National Academy of Sciences, 116(16), 7662–7669.
3. Sinatra, G. M., Kienhues, D., & Hofer, B. K. (2014). Addressing challenges to public understanding of science: Epistemic cognition, motivated reasoning, and conceptual change. Educational Psychologist, 49(2), 123–138.
4. Bonney, R., Phillips, T. B., Ballard, H. L., & Enck, J. W. (2016). Can citizen science enhance public understanding of science? Public Understanding of Science, 25(1), 2–16.
5. Dawson, E. (2014). "Not designed for us": How science museums and science centers socially exclude low-income, minority ethnic groups. Science Education, 98(6), 981–1008.
6. Pandya, R. E. (2012). A framework for engaging diverse communities in citizen science in the US. Frontiers in Ecology and the Environment, 10(6), 314–317.
7. Besley, J. C., Dudo, A., & Yuan, S. (2018). Scientists' views about communication objectives. Public Understanding of Science, 27(6), 708–730.
8. National Research Council. (2012). A framework for K-12 science education: Practices, crosscutting concepts, and core ideas. National Academies Press.
9. Hicks, D. (1995). Published papers, tacit competencies and corporate management of the public/private character of knowledge. Industrial and Corporate Change, 4(2), 401–424.
10. Shirk, J. L., Ballard, H. L., Wilderman, C. C., Phillips, T., Wiggins, A., Jordan, R., ... & Bonney, R. (2012). Public participation in scientific research: A framework for deliberate design. Ecology and Society, 17(2), 29.
11. Sutherland, W. J., Spiegelhalter, D., & Burgman, M. A. (2013). Policy: Twenty tips for interpreting scientific claims. Nature, 503(7476), 335–337.
12. Fischhoff, B. (2019). Evaluating science communication. Proceedings of the National Academy of Sciences, 116(16), 7670–7675.
13. Hartley, B. L., Osborn, F., & Maschner, H. (2020). Broadening participation in STEM: Policy implications of a diverse workforce. Policy Insights from the Behavioral and Brain Sciences, 7(2), 132–139.

A Vision for the Future and a Call to Action

As we look to the future, it is clear that building a scientifically literate society is not just a desirable goal, but an urgent necessity. In a world facing complex challenges such as climate change, global health crises, and rapid technological transformation, the ability to understand and apply scientific knowledge is essential for informed decision-making, effective problem-solving, and responsible citizenship [1]. Moreover, as the pace of scientific discovery and innovation continues to accelerate, the gap between those who have access to and understanding of science and those who do not risks widening, exacerbating existing inequalities and creating new forms of social and economic exclusion [2].

To address these challenges and realize the full potential of science as a driver of human progress and well-being, we need a bold and comprehensive vision for the future of scientific literacy. This vision must be grounded in a commitment to equity, inclusion, and empowerment, recognizing that science is not just a body of knowledge, but a way of thinking and inquiring about the world that should be accessible to all [3]. It must also be driven by a sense of urgency and collective responsibility, acknowledging that building a scientifically literate society is not a task that can be left to others, but one that requires the active engagement and collaboration of individuals, communities, and institutions across all sectors of society [4].

At the core of this vision is a fundamental shift in how we think about and value science education and communication. Rather than treating science as a specialized domain reserved for experts and elites, we need to recognize it as a vital part of our shared cultural heritage and a key enabler of personal and societal well-being [5]. This means moving beyond traditional models of science education that focus on memorizing facts and formulas, and instead emphasizing the development of scientific habits of mind, such as critical thinking, evidence-based reasoning, and creative problem-solving [6]. It also means creating more opportunities for lifelong learning and engagement with science, through formal and informal education, public outreach, and community-based initiatives [7].

To achieve this vision, we need a sustained and coordinated effort to invest in and prioritize science education and communication at all levels of society. This includes providing adequate funding and support for K-12 science education, with a focus on inquiry-based and hands-on learning experiences that foster curiosity, creativity, and analytical skills [8]. It also means strengthening and expanding science education and training programs in higher education, with a particular emphasis on interdisciplinary and applied research that addresses real-world problems and engages diverse stakeholders [9].

Beyond the formal education system, we need to create a more vibrant and inclusive ecosystem for science communication and public engagement. This means supporting and valuing the work of science communicators, journalists, and outreach professionals who play a crucial role in translating scientific knowledge for broader audiences and fostering dialogue and debate about the social and ethical implications of science [10]. It also means leveraging the power of digital media and technologies to create new forms of interactive and participatory science engagement, such as citizen science projects, online learning communities, and immersive science experiences [11].

Importantly, building a scientifically literate society also requires addressing systemic barriers and inequities that limit access to and participation in science. This means confronting and dismantling the historical and ongoing exclusion of marginalized groups from science education and careers, and actively working to create more diverse, inclusive, and equitable scientific communities [12]. It also means recognizing and valuing the knowledge and perspectives of indigenous and local communities, and engaging them as partners and leaders in scientific research and decision-making [13].

Ultimately, the vision for a scientifically literate society is one in which science is not just a subject to be studied, but a way of thinking and being in the world that is integrated into all aspects of personal, social, and political life. It is a vision in which every individual has the knowledge, skills, and confidence to engage with science in meaningful and empowering ways, and in which scientific knowledge is used to inform and enrich public discourse, policy-making, and collective action [14].

Realizing this vision will not be easy, and it will require significant changes in how we organize and prioritize science education, communication, and engagement. But the stakes could not be higher. In a world facing unprecedented challenges and opportunities, the ability to harness the power of science for the greater good is not just a luxury, but a necessity. It is a moral and practical imperative that demands the attention and action of us all [15].

So let this be a call to action – to individuals, communities, institutions, and leaders across all sectors of society. Let us come together to champion the value of science and to work towards a future in which scientific literacy is a fundamental right and a shared responsibility. Let us invest in and support science education and communication, not just as a means to an end, but as an essential part of what it means to be human in the 21st century. And let us do so with a sense of urgency, creativity, and resolve, knowing that the choices we make today will shape the world we leave to future generations.

The path ahead may be challenging, but it is also filled with possibility and promise. By embracing science as a vital part of our individual and collective lives, we open up new horizons of understanding, discovery, and action. We empower ourselves and each other to ask better questions, to seek better answers, and to create better solutions to the challenges we face. And we reaffirm our shared commitment to the pursuit of knowledge, the power of reason, and the potential of human ingenuity to transform the world for the better.

So let us begin this journey together, with open minds, caring hearts, and a fierce determination to build a scientifically literate society that is worthy of the dreams and aspirations of all who inhabit this precious planet we call home. The future is ours to shape, and the time to act is now.

References

1. National Academies of Sciences, Engineering, and Medicine. (2016). Science literacy: Concepts, contexts, and consequences. National Academies Press.
2. Feinstein, N. W., Allen, S., & Jenkins, E. (2013). Outside the pipeline: Reimagining science education for nonscientists. Science, 340(6130), 314–317.
3. Bang, M., & Medin, D. (2010). Cultural processes in science education: Supporting the navigation of multiple epistemologies. Science Education, 94(6), 1008–1026.
4. Tate, E. D. (2012). Science literacy and the pursuit of a democratic society. Journal of Science Teacher Education, 23(7), 645–663.
5. Hurd, P. D. (1998). Scientific literacy: New minds for a changing world. Science Education, 82(3), 407–416.
6. National Research Council. (2012). A framework for K-12 science education: Practices, crosscutting concepts, and core ideas. National Academies Press.
7. Falk, J. H., & Dierking, L. D. (2010). The 95 percent solution: School is not where most Americans learn most of their science. American Scientist, 98(6), 486–493.
8. Osborne, J. (2014). Teaching scientific practices: Meeting the challenge of change. Journal of Science Teacher Education, 25(2), 177–196.

9. Crow, M. M., & Dabars, W. B. (2015). Designing the new American university. JHU Press.
10. Besley, J. C., & Tanner, A. H. (2011). What science communication scholars think about training scientists to communicate. Science Communication, 33(2), 239–263.
11. Bonney, R., Phillips, T. B., Ballard, H. L., & Enck, J. W. (2016). Can citizen science enhance public understanding of science? Public Understanding of Science, 25(1), 2–16.
12. Medin, D. L., & Lee, C. D. (2012). Diversity makes better science. APS Observer, 25(5).
13. David-Chavez, D. M., & Gavin, M. C. (2018). A global assessment of Indigenous community engagement in climate research. Environmental Research Letters, 13(12), 123005.
14. Roberts, D. A. (2007). Scientific literacy/science literacy. In S. K. Abell & N. G. Lederman (Eds.), Handbook of research on science education (pp. 729–780). Lawrence Erlbaum Associates.
15. Fischhoff, B. (2019). Evaluating science communication. Proceedings of the National Academy of Sciences, 116(16), 7670–7675.

Conclusion

Recap of the Book's Main Points

Throughout this book, we have explored the critical importance of scientific literacy in the modern world and the urgent need to address the current state of scientific illiteracy in the United States. We began by examining the extent of the problem, citing statistics that reveal a troubling lack of basic scientific knowledge among the American public, as well as a significant gap between the scientific literacy of the U.S. and that of many other developed nations [1]. We also investigated the historical trends and factors that have contributed to this situation, including inadequacies in K-12 science education, the politicization of science, and the spread of misinformation through various media channels [2].

The consequences of scientific illiteracy, as we have seen, are far-reaching and profound. In the realm of public health and personal well-being, misconceptions about diseases, vaccines, and treatments can lead to poor health choices, delayed medical care, and the spread of preventable illnesses [3]. A lack of understanding about environmental science and climate change can hinder individual actions and support for policies aimed at mitigating the devastating effects of global warming [4]. In the realm of technological advancement and economic competitiveness, a scientifically illiterate workforce is ill-equipped to drive innovation and maintain the nation's position as a global leader in science and technology [5].

Moreover, scientific illiteracy poses significant challenges to political decision-making and the functioning of democracy itself. When large segments of the population lack the knowledge and critical thinking skills to evaluate scientific evidence, they are more susceptible to misinformation and manipulation, which can skew voter choices and degrade the quality of political discourse [6]. This, in turn, can lead to policies that are not grounded in scientific reality, with potentially harmful consequences for public health, environmental sustainability, and social progress [7].

Addressing these challenges, as we have discussed, will require a multi-faceted approach that involves educators, policymakers, scientists, and the broader public. In the realm of education, strategies for enhancing K-12 science curricula and teaching methods include emphasizing hands-on, inquiry-based learning, integrating real-world applications, and leveraging technology to support personalized and adaptive learning [8]. Universities and informal science education programs, such as museums and science centers, also have a vital role to play in promoting scientific literacy and engagement across all ages and backgrounds [9].

Equally important is the need to improve science communication and public outreach. By developing clear, compelling, and accessible ways of communicating scientific information, scientists, journalists, and institutions can help to bridge the gap between the scientific community and the general public [10]. Successful examples of public engagement with science, such as citizen science projects and science festivals, demonstrate the power of participatory and experiential learning to foster curiosity, understanding, and appreciation for scientific research [11].

Ultimately, building a scientifically literate society is not just a matter of imparting knowledge, but of cultivating a culture that values science as a way of understanding the world and making informed decisions. The benefits of a scientifically literate population are manifold, from improved health outcomes and environmental stewardship to increased innovation and economic competitiveness [12]. Achieving this vision will require the concerted efforts of individuals, communities, and policymakers alike, working together to prioritize science education, communication, and engagement as essential components of a thriving democracy [13].

The path forward, as we have acknowledged, is not without its challenges. Overcoming the deep-seated problems of scientific illiteracy will require significant investments of time, resources, and political will. It will also require a willingness to confront uncomfortable truths about the ways in which science has been misused or misrepresented in the past, and a commitment to building a more equitable and inclusive scientific enterprise [14].

Yet, as we have argued throughout this book, the stakes could not be higher. In a world facing complex and urgent challenges, from climate change and infectious diseases to artificial intelligence and space exploration, the ability to understand and apply scientific knowledge is not just a luxury, but a necessity. It is a fundamental prerequisite for informed citizenship, effective problem-solving, and responsible stewardship of our planet and our future [15].

By highlighting the extent and consequences of scientific illiteracy, exploring strategies for improving science education and communication, and envisioning a future in which science is a vital part of our shared cultural heritage, this book has sought to contribute to the ongoing conversation about the role of science in society. It is a conversation that concerns us all, and one that will shape the course of human progress for generations to come.

Ultimately, the message of this book is one of hope and empowerment. By working together to build a scientifically literate society, we have the opportunity to unlock the full potential of science as a tool for understanding, exploring, and improving the world around us. It is an opportunity that we cannot afford to miss, and one that will require the dedication and collaboration of us all.

References

1. National Science Board. (2018). Science and Engineering Indicators 2018. National Science Foundation.
2. Oreskes, N., & Conway, E. M. (2010). Merchants of doubt: How a handful of scientists obscured the truth on issues from tobacco smoke to global warming. Bloomsbury Press.
3. Larson, H. J., Jarrett, C., Eckersberger, E., Smith, D. M., & Paterson, P. (2014). Understanding vaccine hesitancy around vaccines and vaccination from a global perspective: a systematic review of published literature, 2007–2012. Vaccine, 32(19), 2150-2159.
4. Hornsey, M. J., Harris, E. A., Bain, P. G., & Fielding, K. S. (2016). Meta-analyses of the determinants and outcomes of belief in climate change. Nature Climate Change, 6(6), 622-626.
5. National Research Council. (2012). Rising above the gathering storm: Developing regional innovation environments: A workshop summary. National Academies Press.
6. Scheufele, D. A., & Krause, N. M. (2019). Science audiences, misinformation, and fake news. Proceedings of the National Academy of Sciences, 116(16), 7662-7669.
7. Dietz, T. (2013). Bringing values and deliberation to science communication. Proceedings of the National Academy of Sciences, 110(Supplement 3), 14081-14087.
8. Freeman, S., Eddy, S. L., McDonough, M., Smith, M. K., Okoroafor, N., Jordt, H., & Wenderoth, M. P. (2014). Active learning increases student performance in science, engineering, and mathematics. Proceedings of the National Academy of Sciences, 111(23), 8410-8415.

9. Bell, P., Lewenstein, B., Shouse, A. W., & Feder, M. A. (Eds.). (2009). Learning science in informal environments: People, places, and pursuits. National Academies Press.
10. Nisbet, M. C., & Scheufele, D. A. (2009). What's next for science communication? Promising directions and lingering distractions. American Journal of Botany, 96(10), 1767-1778.
11. Bonney, R., Phillips, T. B., Ballard, H. L., & Enck, J. W. (2016). Can citizen science enhance public understanding of science? Public Understanding of Science, 25(1), 2-16.
12. National Academies of Sciences, Engineering, and Medicine. (2016). Science literacy: Concepts, contexts, and consequences. National Academies Press.
13. Feinstein, N. W., & Meshoulam, D. (2014). Science for what public? Addressing equity in American science museums and science centers. Journal of Research in Science Teaching, 51(3), 368-394.
14. Berhe, A. A., Barnes, R. T., Hastings, M. G., Mattheis, A., Schneider, B., Williams, B. M., & Marín-Spiotta, E. (2022). Scientists from historically excluded groups face a hostile obstacle course. Nature Geoscience, 15(1), 2-4.
15. Hicks, D. (2017). Scientific controversies as proxy politics. Science, Technology, & Human Values, 42(6), 1039-1065.

The Urgency of Addressing Scientific Illiteracy in America

As we conclude this exploration of scientific literacy and its vital importance in the modern world, it is imperative that we underscore the urgency of addressing the current state of scientific illiteracy in the United States. The consequences of a society that lacks a fundamental understanding of scientific principles and methods are not merely academic or hypothetical; they are real, tangible, and increasingly dire in the face of the complex challenges we face as a nation and as a global community.

In a world grappling with the existential threat of climate change, the devastating impacts of global pandemics, and the breakneck pace of technological transformation, the ability to comprehend and apply scientific knowledge is not just a luxury, but a necessity. It is the foundation upon which we must build our collective capacity to make informed decisions, develop innovative solutions, and chart a course towards a more sustainable, equitable, and prosperous future [1].

Yet, as we have seen throughout this book, the United States is currently facing a crisis of scientific literacy that threatens to undermine our ability to meet these challenges head-on. From the persistence of misinformation and pseudoscience to the erosion of public trust in scientific institutions, the warning signs are clear and compelling. If we fail to take swift and decisive action to address this crisis, we risk not only falling behind our global competitors in the realm of scientific and technological advancement but also jeopardizing the very foundations of our democracy and our way of life [2].

The COVID-19 pandemic has brought this urgency into sharp relief, exposing the devastating consequences of scientific illiteracy on a massive scale. From the proliferation of conspiracy theories and the rejection of evidence-based public health measures to the unequal distribution of the health and economic burdens of the virus, the pandemic has laid bare the fault lines of a society ill-equipped to navigate the complexities of a scientific crisis [3]. It has also highlighted the critical importance of effective science communication and public engagement in building trust, countering misinformation, and mobilizing collective action in the face of a shared threat [4].

But the implications of scientific illiteracy extend far beyond the realm of public health. In the fight against climate change, for example, a lack of understanding about the basic science of global warming and its potential impacts can lead to a lack of support for the policies and actions needed to mitigate its worst effects [5]. Similarly, in the realm of technology, a scientifically illiterate population is more vulnerable to the risks and unintended consequences of rapidly advancing fields such as artificial intelligence, biotechnology, and digital surveillance, while also being less equipped to harness their potential benefits for society [6].

Moreover, scientific illiteracy is not just a matter of individual knowledge and understanding; it is also a deeply social and political problem that reflects and reinforces broader inequities and power imbalances within our society. From the unequal access to quality science education and resources across different communities to the historical exclusion and marginalization of certain groups within the scientific enterprise itself, the roots of scientific illiteracy are inextricably linked to larger structures of privilege, oppression, and injustice [7].

Addressing these challenges will require a sustained and multi-faceted effort that engages all sectors of society, from education and research to policy and public engagement. It will require significant investments in science education and communication, as well as a fundamental rethinking of how we value and prioritize scientific knowledge and expertise within our culture and our institutions [8].

It will also require a willingness to confront uncomfortable truths about the ways in which science has been misused or misrepresented in the past, and a commitment to building a more inclusive, equitable, and accountable scientific enterprise moving forward. This means not only working to diversify the scientific workforce and leadership but also engaging communities that have been historically marginalized or excluded from the scientific process as full partners and co-creators of knowledge [9].

Ultimately, the urgency of addressing scientific illiteracy in America stems from the recognition that science is not just a collection of facts and figures, but a powerful tool for understanding and shaping the world around us. It is a way of asking questions, seeking evidence, and testing hypotheses that has the potential to unlock new frontiers of discovery, innovation, and human progress. But it is also a deeply human endeavor, one that reflects our values, our priorities, and our aspirations as a society [10].

By investing in scientific literacy and embracing science as a vital part of our shared cultural heritage, we have the opportunity to empower individuals and communities to be active participants in the scientific enterprise, to hold those in power accountable for evidence-based decision-making, and to work together towards a more just, sustainable, and flourishing world.

The stakes could not be higher, and the time for action is now. As we look to the future, let us summon the courage, the creativity, and the collective will to rise to this challenge, and to build a society that values and upholds the transformative power of science for the benefit of all. The path ahead may be difficult, but the cost of inaction is far too great to bear. It is up to us to seize this moment and to chart a new course towards a brighter, more scientifically literate future for ourselves and for generations to come.

References

1. National Academies of Sciences, Engineering, and Medicine. (2018). The integration of the humanities and arts with sciences, engineering, and medicine in higher education: Branches from the same tree. National Academies Press.
2. Oreskes, N. (2019). Why trust science? Princeton University Press.
3. Germani, F., & Biller-Andorno, N. (2021). The anti-vaccination infodemic on social media: A behavioral analysis. PLoS One, 16(3), e0247642.
4. Kreps, S. E., & Kriner, D. L. (2020). Model uncertainty, political contestation, and public trust in science: Evidence from the COVID-19 pandemic. Science Advances, 6(43), eabd4563.
5. Van der Linden, S., Leiserowitz, A., Rosenthal, S., & Maibach, E. (2017). Inoculating the public against misinformation about climate change. Global Challenges, 1(2), 1600008.
6. Sinatra, G. M., & Hofer, B. K. (2016). Public understanding of science: Policy and educational implications. Policy Insights from the Behavioral and Brain Sciences, 3(2), 245-253.
7. Berhe, A. A., Barnes, R. T., Hastings, M. G., Mattheis, A., Schneider, B., Williams, B. M., & Marín-Spiotta, E. (2022). Scientists from historically excluded groups face a hostile obstacle course. Nature Geoscience, 15(1), 2-4.
8. Feinstein, N. W., & Meshoulam, D. (2014). Science for what public? Addressing equity in American science museums and science centers. Journal of Research in Science Teaching, 51(3), 368-394.
9. Canfield, K. N., Menezes, S., Matsuda, S. B., Moore, A., Mosley Austin, A. N., Dewsbury, B. M., ... & Taylor, C. (2020). Science communication demands a critical approach that centers inclusion, equity, and intersectionality. Frontiers in Communication, 5, 2.
10. Hicks, D. (2017). Scientific controversies as proxy politics. Science, Technology, & Human Values, 42(6), 1039-1065.

The Potential for Positive Change and Progress

While the current state of scientific illiteracy in the United States presents a daunting challenge, it is important to recognize that this crisis also represents an opportunity for profound positive change and progress. By confronting the root causes of scientific illiteracy and investing in the strategies and solutions outlined in this book, we have the potential to not only improve scientific understanding among the general public but also to catalyze broader shifts in our society towards greater equity, inclusion, and evidence-based decision-making.

At the heart of this potential lies the transformative power of education. As we have discussed, improving science education at all levels, from K-12 schools to universities and informal learning environments, is essential for cultivating a more scientifically literate populace. But the benefits of this investment extend far beyond the realm of science itself. By equipping students with the critical thinking skills, problem-solving abilities, and intellectual curiosity that are the hallmarks of scientific inquiry, we can empower them to become lifelong learners, engaged citizens, and innovative leaders in all fields [1].

Moreover, by reimagining science education in ways that prioritize hands-on experience, real-world applications, and cross-disciplinary connections, we can help to break down the silos that have traditionally separated science from other domains of human knowledge and endeavor. This more holistic and integrative approach to education has the potential to foster new forms of creativity, collaboration, and innovation that can drive progress across a wide range of social, economic, and environmental challenges [2].

In addition to education, improving scientific literacy also requires a fundamental shift in how we communicate and engage with the public about science. As we have seen, the traditional model of science communication, in which experts simply convey information to passive audiences, is no longer sufficient in an age of social media, misinformation, and declining trust in institutions. Instead, we need to embrace a more dialogic and participatory approach to science communication, one that invites the public to be active partners in the scientific process and that values diverse forms of knowledge and expertise [3].

By creating more opportunities for scientists to collaborate with communities, journalists, policymakers, and other stakeholders, we can help to build trust, understanding, and a shared sense of ownership over the scientific enterprise. This more inclusive and responsive approach to science communication has the potential to not only improve public understanding of science but also to enhance the relevance and impact of scientific research itself. By listening to and learning from the needs, concerns, and insights of diverse publics, scientists can ask better questions, design more effective solutions, and contribute to more equitable and sustainable outcomes for society [4].

Ultimately, addressing scientific illiteracy is not just about improving individual knowledge and understanding; it is also about creating a more just, humane, and democratic society. By embracing science as a vital tool for informing public policy and decision-making, we can help to ensure that our collective choices are grounded in evidence, reason, and a commitment to the greater good. This means not only investing in scientific research and innovation but also in the social, political, and ethical frameworks that govern how science is used and applied in service of human well-being [5].

It also means grappling honestly with the ways in which science has been misused or abused in the past, and working to build a more accountable and ethically grounded scientific culture moving forward. This requires a willingness to confront and dismantle the systemic inequities and biases that have long plagued the scientific enterprise, from the underrepresentation of women and people of color in STEM fields to the exploitation of marginalized communities in scientific research. By actively working to create a more diverse, inclusive, and socially responsible scientific community, we can help to ensure that the benefits of scientific progress are shared more equitably across society [6].

Of course, none of this will be easy. Overcoming the deeply entrenched challenges of scientific illiteracy will require sustained effort, resources, and political will over many years and decades. It will also require a willingness to think creatively and expansively about the role of science in our lives and in our world, and to embrace new forms of collaboration, experimentation, and public engagement that may push us beyond our comfort zones.

But the potential rewards of this work are immense. By cultivating a more scientifically literate and engaged society, we can unlock new frontiers of discovery and innovation, from the depths of the oceans to the farthest reaches of space. We can harness the power of science to solve the most pressing challenges facing our planet, from climate change and public health crises to food insecurity and social inequality. And we can create a more resilient, adaptable, and hopeful society, one that is better equipped to navigate the complex and rapidly changing world of the 21st century [7].

Ultimately, the choice is ours. We can continue down the path of scientific illiteracy and disengagement, and face the consequences of a society increasingly divided, misinformed, and ill-prepared for the challenges ahead. Or we can choose to invest in the power of science and education to transform our world for the better, and to create a brighter, more scientifically literate future for ourselves and for generations to come.

The potential is there. The tools and strategies are within our grasp. What is needed now is the courage, the commitment, and the collective will to act. Let us seize this moment, and work together to build a society that values and upholds the vital role of science in advancing human knowledge, well-being, and progress. The future is ours to shape, and the time to begin is now.

References

1. National Research Council. (2012). Education for life and work: Developing transferable knowledge and skills in the 21st century. National Academies Press.
2. National Academies of Sciences, Engineering, and Medicine. (2018). The integration of the humanities and arts with sciences, engineering, and medicine in higher education: Branches from the same tree. National Academies Press.
3. Dudo, A., & Besley, J. C. (2016). Scientists' prioritization of communication objectives for public engagement. PLoS One, 11(2), e0148867.
4. Safford, H. D., Sawyer, S. C., Kocher, S. D., Hiers, J. K., & Cross, M. (2017). Linking knowledge to action: The role of boundary spanners in translating ecology. Frontiers in Ecology and the Environment, 15(10), 560-568.
5. Jasanoff, S. (2012). Science and public reason. Routledge.
6. Hofstra, B., Kulkarni, V. V., Galvez, S. M. N., He, B., Jurafsky, D., & McFarland, D. A. (2020). The diversity–innovation paradox in science. Proceedings of the National Academy of Sciences, 117(17), 9284-9291.
7. National Academies of Sciences, Engineering, and Medicine. (2020). Promising practices for addressing the underrepresentation of women in science, engineering, and medicine: Opening doors. National Academies Press.

Resources for Further Reading and Engagement

As we conclude our exploration of scientific literacy and its vital importance in the modern world, it is important to recognize that this book is just the beginning of a much larger conversation. For those readers who have been inspired by the ideas and arguments presented here, and who wish to deepen their understanding and engagement with these issues, there is a wealth of additional resources available.

One of the best ways to continue learning about scientific literacy is to explore the work of leading scholars, educators, and science communicators who have made significant contributions to this field. For example, the writings of Neil deGrasse Tyson [1], Michio Kaku [2], and Carl Sagan [3] offer compelling and accessible introductions to the wonders of science and the importance of scientific thinking in our daily lives. Similarly, the work of science education researchers such as Judith Lederman [4] and Jonathan Osborne [5] provides valuable insights into the challenges and opportunities of improving science education in schools and beyond.

Another valuable resource for further reading is the growing body of literature on science communication and public engagement with science. Books such as "The Oxford Handbook of the Science of Science Communication" [6] and "Communicating Science Effectively: A Research Agenda" [7] offer comprehensive overviews of the latest research and best practices in this field, while works like "The Rightful Place of Science: Citizen Science" [8] and "The Science of Science Communication II: Summary of a Colloquium" [9] explore specific strategies for involving the public in scientific research and decision-making.

For those interested in the broader social, political, and ethical dimensions of science and technology, there are many excellent resources available. Books such as "The Fifth Risk" by Michael Lewis [10] and "The Death of Expertise" by Tom Nichols [11] offer thoughtful and provocative analyses of the current state of public trust in science and expertise, while works like "The Politics of Evidence-Based Policy Making" by Paul Cairney [12] and "The Honest Broker" by Roger Pielke Jr. [13] explore the complex relationships between science, policy, and politics.

In addition to books and articles, there are also many online resources and communities that can help individuals stay informed and engaged with issues related to scientific literacy and public engagement with science. Websites such as ScienceDaily [14], Phys.org [15], and Science Friday [16] offer a steady stream of news and commentary on the latest scientific discoveries and controversies, while blogs like Science-Based Medicine [17] and Science-Based Life [18] provide critical perspectives on pseudoscience and misinformation.

Social media platforms such as Twitter and Facebook can also be valuable tools for connecting with scientists, science communicators, and other advocates for scientific literacy. By following accounts such as @ScienceNewsOrg [19], @AAAS [20], @ScienceAlert [21], and @ScienceOnReddit [22], individuals can stay up-to-date on the latest developments and participate in ongoing conversations about science and society.

For those who prefer more interactive and experiential forms of learning, there are also many opportunities to engage with science through museums, science centers, and other informal education settings. Institutions such as the Smithsonian National Museum of Natural History [23], the California Academy of Sciences [24], and the American Museum of Natural History [25] offer a wide range of exhibits, programs, and resources designed to engage visitors of all ages and backgrounds with the wonders of science.

Citizen science projects, in which members of the public participate in real scientific research, are another excellent way to gain hands-on experience with the scientific process. Websites such as Zooniverse [26], SciStarter [27], and CitizenScience.gov [28] provide access to a wide range of projects spanning fields from astronomy and ecology to public health and social science.

Ultimately, the key to promoting scientific literacy and public engagement with science is to create a culture of lifelong learning and curiosity. By seeking out new sources of knowledge and experience, asking questions, and engaging in dialogue with others who share our interests and concerns, we can all become active participants in the ongoing quest to understand and shape the world around us.

Of course, this is not always easy. In a world filled with competing demands on our time and attention, it can be tempting to tune out the noise and retreat into our own bubbles of comfort and familiarity. But as the challenges facing our society grow ever more complex and urgent, the need for a scientifically literate and engaged citizenry has never been greater.

By making a commitment to ongoing learning and engagement, and by working together to build a more informed, empowered, and inclusive society, we can all play a vital role in shaping a brighter, more scientifically literate future. The resources and communities highlighted here are just a starting point – the rest is up to us.

References

1. Tyson, N. D. (2017). Astrophysics for people in a hurry. W. W. Norton & Company.
2. Kaku, M. (2014). The future of the mind: The scientific quest to understand, enhance, and empower the mind. Doubleday.
3. Sagan, C. (1996). The demon-haunted world: Science as a candle in the dark. Random House.
4. Lederman, J. S. (2019). Teaching and learning nature of scientific knowledge: Is it déjà vu all over again? Disciplinary and Interdisciplinary Science Education Research, 1(1), 1-9.
5. Osborne, J. (2014). Teaching scientific practices: Meeting the challenge of change. Journal of Science Teacher Education, 25(2), 177-196.
6. Jamieson, K. H., Kahan, D., & Scheufele, D. A. (Eds.). (2017). The Oxford handbook of the science of science communication. Oxford University Press.
7. National Academies of Sciences, Engineering, and Medicine. (2017). Communicating science effectively: A research agenda. National Academies Press.
8. Cavalier, D., & Kennedy, E. B. (Eds.). (2016). The rightful place of science: Citizen science. Consortium for Science, Policy & Outcomes.
9. National Academies of Sciences, Engineering, and Medicine. (2014). The science of science communication II: Summary of a colloquium. National Academies Press.
10. Lewis, M. (2018). The fifth risk. W. W. Norton & Company.
11. Nichols, T. (2017). The death of expertise: The campaign against established knowledge and why it matters. Oxford University Press.
12. Cairney, P. (2016). The politics of evidence-based policy making. Springer.
13. Pielke, R. A. (2007). The honest broker: Making sense of science in policy and politics. Cambridge University Press.
14. ScienceDaily. (n.d.). ScienceDaily: Your source for the latest research news. https://www.sciencedaily.com/
15. Phys.org. (n.d.). Phys.org–News and articles on science and technology. https://phys.org/
16. Science Friday. (n.d.). Science Friday. https://www.sciencefriday.com/
17. Science-Based Medicine. (n.d.). Science-Based Medicine. https://sciencebasedmedicine.org/
18. Science-Based Life. (n.d.). Science-Based Life. https://sciencebasedlife.wordpress.com/
19. Science News [@ScienceNewsOrg]. (n.d.). Science News on Twitter. https://twitter.com/ScienceNewsOrg
20. AAAS [@AAAS]. (n.d.). AAAS on Twitter. https://twitter.com/AAAS
21. ScienceAlert [@ScienceAlert]. (n.d.). ScienceAlert on Twitter. https://twitter.com/ScienceAlert
22. Science on Reddit [@ScienceOnReddit]. (n.d.). Science on Reddit on Twitter. https://twitter.com/ScienceOnReddit

23. Smithsonian National Museum of Natural History. (n.d.). Smithsonian National Museum of Natural History. https://naturalhistory.si.edu/
24. California Academy of Sciences. (n.d.). California Academy of Sciences. https://www.calacademy.org/
25. American Museum of Natural History. (n.d.). American Museum of Natural History. https://www.amnh.org/
26. Zooniverse. (n.d.). Zooniverse. https://www.zooniverse.org/
27. SciStarter. (n.d.). SciStarter: Science we can do together. https://scistarter.org/
28. CitizenScience.gov. (n.d.). CitizenScience.gov. https://www.citizenscience.gov/

Glossary of Key Scientific Terms and Concepts

As we have explored throughout this book, scientific literacy is not just about memorizing facts and figures, but about understanding the fundamental concepts, methods, and practices that underlie scientific inquiry. To help readers navigate the complex terrain of scientific knowledge, we have compiled this glossary of key terms and concepts that are essential for engaging with science in a meaningful way.

Empirical evidence: Empirical evidence refers to information or data that is obtained through observation, experimentation, or measurement [1]. It is the foundation of the scientific method, which holds that knowledge about the natural world should be based on verifiable evidence rather than opinion, authority, or intuition. Empirical evidence can take many forms, from the results of controlled experiments to data gathered through surveys or field observations.

Hypothesis: A hypothesis is a proposed explanation for a phenomenon or a tentative answer to a research question [2]. In the scientific method, hypotheses are used to guide the design of experiments or studies, which seek to gather evidence that can support, refute, or modify the initial hypothesis. A good hypothesis is testable, meaning that it can be evaluated using empirical evidence, and falsifiable, meaning that it is possible to imagine evidence that would disprove it [3].

Peer review: Peer review is the process by which scientific research is evaluated by other experts in the same field before it is published or disseminated [4]. The goal of peer review is to ensure that research is rigorous, reliable, and significant, and to catch any errors or weaknesses before they become part of the scientific record. Peer reviewers are typically anonymous and are chosen for their expertise and impartiality. While peer review is not perfect, it is widely regarded as a crucial quality control mechanism in science [5].

Replication: Replication refers to the process of repeating a scientific study or experiment to see if the same results can be obtained [6]. Replication is a cornerstone of the scientific method, as it helps to ensure that findings are robust and generalizable, rather than the result of chance or error. When a study is successfully replicated, it increases confidence in the validity of the original findings. Conversely, when a study fails to replicate, it suggests that the original findings may have been flawed or that important variables were overlooked [7].

Sample size: Sample size refers to the number of individuals or observations included in a scientific study or experiment [8]. In general, larger sample sizes are preferred because they provide more reliable and representative results. However, sample size must be balanced against practical considerations such as cost, time, and feasibility. Statistical methods can be used to determine the minimum sample size needed to detect a given effect or relationship [9].

Correlation: Correlation refers to the degree to which two variables are related or associated with each other [10]. A positive correlation means that as one variable increases, the other tends to increase as well. A negative correlation means that as one variable increases, the other tends to decrease. Importantly, correlation does not necessarily imply causation – just because two variables are correlated does not mean that one causes the other [11]. Determining causality requires additional evidence, such as experimental manipulation or theoretical reasoning.

Statistical significance: Statistical significance refers to the likelihood that a result or relationship observed in a study is due to chance rather than a real effect [12]. Researchers use statistical tests to calculate the probability (p-value) of obtaining their results if there were no real effect. By convention, a p-value of less than 0.05 (5%) is considered statistically significant, meaning that there is a less than 5% chance that the results are due to chance alone. However, statistical significance does not necessarily imply practical or clinical significance, and must be interpreted in the context of other factors such as effect size and study design [13].

Systematic review: A systematic review is a type of study that uses a structured and transparent methodology to identify, select, and synthesize all of the available research evidence on a given topic [14]. Systematic reviews are considered the highest level of evidence in many fields, as they provide a comprehensive and unbiased assessment of the state of knowledge on a particular question. Systematic reviews often include a meta-analysis, which is a statistical technique for combining the results of multiple studies to obtain a more precise estimate of the overall effect or relationship [15].

Randomized controlled trial: A randomized controlled trial (RCT) is a type of experimental study design in which participants are randomly assigned to receive either an intervention (such as a new drug or treatment) or a control (such as a placebo or standard care) [16]. RCTs are considered the gold standard for evaluating the effectiveness of interventions, as they help to minimize bias and confounding variables. By randomly assigning participants to groups, researchers can ensure that any differences in outcomes are due to the intervention itself rather than other factors [17].

Scientific consensus: Scientific consensus refers to the collective judgment or agreement of the scientific community on a particular issue, based on the available evidence [18]. Scientific consensus is typically established through a process of ongoing research, debate, and peer review, in which multiple lines of evidence converge to support a particular conclusion. Examples of scientific consensus include the existence of climate change, the safety and effectiveness of vaccines, and the theory of evolution by natural selection. While scientific consensus is not infallible, it represents the best available understanding of a given topic based on the current state of knowledge [19].

By familiarizing themselves with these and other key scientific concepts, readers can become more confident and capable consumers of scientific information. At the same time, it is important to remember that scientific knowledge is always evolving, and that there is much that we still do not know or understand. The goal of scientific literacy is not to have all the answers, but to ask better questions, to think critically about evidence, and to engage in ongoing learning and discovery. With these tools and mindsets, we can all participate in the exciting and transformative enterprise of science, and help to build a more informed, engaged, and scientifically literate society.

References

1. Haig, B. D. (2005). An abductive theory of scientific method. Psychological Methods, 10(4), 371-388.
2. Kerlinger, F. N., & Lee, H. B. (2000). Foundations of behavioral research (4th ed.). Harcourt College Publishers.
3. Popper, K. R. (1959). The logic of scientific discovery. Hutchinson.
4. Bornmann, L. (2011). Scientific peer review. Annual Review of Information Science and Technology, 45(1), 197-245.
5. Smith, R. (2006). Peer review: A flawed process at the heart of science and journals. Journal of the Royal Society of Medicine, 99(4), 178-182.
6. Schmidt, S. (2009). Shall we really do it again? The powerful concept of replication is neglected in the social sciences. Review of General Psychology, 13(2), 90-100.
7. Open Science Collaboration. (2015). Estimating the reproducibility of psychological science. Science, 349(6251), aac4716.
8. Lenth, R. V. (2001). Some practical guidelines for effective sample size determination. The American Statistician, 55(3), 187-193.
9. Button, K. S., Ioannidis, J. P., Mokrysz, C., Nosek, B. A., Flint, J., Robinson, E. S., & Munafò, M. R. (2013). Power failure: Why small sample size undermines the reliability of neuroscience. Nature Reviews Neuroscience, 14(5), 365-376.
10. Goodwin, C. J. (2010). Research in psychology: Methods and design (6th ed.). John Wiley & Sons.
11. Aldrich, J. (1995). Correlations genuine and spurious in Pearson and Yule. Statistical Science, 10(4), 364-376.
12. Wasserstein, R. L., & Lazar, N. A. (2016). The ASA statement on p-values: Context, process, and purpose. The American Statistician, 70(2), 129-133.
13. Greenland, S., Senn, S. J., Rothman, K. J., Carlin, J. B., Poole, C., Goodman, S. N., & Altman, D. G. (2016). Statistical tests, P values, confidence intervals, and power: A guide to misinterpretations. European Journal of Epidemiology, 31(4), 337-350.
14. Petticrew, M., & Roberts, H. (2006). Systematic reviews in the social sciences: A practical guide. Blackwell Publishing.
15. Borenstein, M., Hedges, L. V., Higgins, J. P., & Rothstein, H. R. (2011). Introduction to meta-analysis. John Wiley & Sons.
16. Schulz, K. F., Altman, D. G., & Moher, D. (2010). CONSORT 2010 statement: Updated guidelines for reporting parallel group randomised trials. BMJ, 340, c332.
17. Sibbald, B., & Roland, M. (1998). Understanding controlled trials: Why are randomised controlled trials important? BMJ, 316(7126), 201.

18. Oreskes, N. (2004). The scientific consensus on climate change. Science, 306(5702), 1686-1686.
19. Cook, J., Oreskes, N., Doran, P. T., Anderegg, W. R., Verheggen, B., Maibach, E. W., ... & Nuccitelli, D. (2016). Consensus on consensus: A synthesis of consensus estimates on human-caused global warming. Environmental Research Letters, 11(4), 048002.

www.ingramcontent.com/pod-product-compliance
Lightning Source LLC
Chambersburg PA
CBHW071504220526
45472CB00003B/912